THE CANCER CODE

ALSO BY JASON FUNG

Life in the Fasting Lane

The Diabetes Code

The Complete Guide to Fasting

The Obesity Code

THE CANCER CODE

Understanding Cancer
As an Evolutionary Disease

DR. JASON FUNG

HARPER

NEW YORK • LONDON • TORONTO • SYDNEY

HARPER

This book contains advice and information relating to health care. It should be used to supplement rather than replace the advice of your doctor or another trained health professional. If you know or suspect you have a health problem, it is recommended that you seek your physician's advice before embarking on any medical program or treatment. All efforts have been made to assure the accuracy of the information contained in this book as of the date of publication. This publisher and the author disclaim liability for any medical outcomes that may occur as a result of applying the methods suggested in this book.

A hardcover edition of this book was published in 2023 by Harper Wave, an imprint of HarperCollins Publishers.

HarperCollins books may be purchased for educational, business, or sales promotional use. For information, please email the Special Markets Department at SPsales@harpercollins.com.

Unless otherwise noted, illustrations are courtesy of the author:
p. 50, Figure 4.1: Bertrand Jordan; p. 170, Figure 13.1: Dr. David Goode;
p. 272 & p. 277, Figure 21.5 & 21.9: Dr. Mette Kalager

FIRST HARPER PAPERBACKS PUBLISHED 2024.

Designed by Bonni Leon-Berman

The Library of Congress has catalogued the hardcover edition as follows:

Names: Fung, Jason, author.
Title: The cancer code : a revolutionary new understanding of a medical mystery / Jason Fung.
Identifiers: LCCN 2020019321 (print) | LCCN 2020019322 (ebook) | ISBN 9780062894007 (hardcover) | ISBN 9780062894021 (epub)
Subjects: MESH: Neoplasms
Classification: LCC RC263 (print) | LCC RC263 (ebook) | NLM QZ 200 | DDC 616.99/4—dc23
LC record available at https://lccn.loc.gov/2020019321
LC ebook record available at https://lccn.loc.gov/2020019322

ISBN 978-0-06-289401-4 (pbk.)

24 25 26 27 28 LBC 5 4 3 2 1

Dedicated to my beautiful wife, Mina, and my sons, Jonathan and Matthew, for all their love, support, and patience. I could not have done it without you.

CONTENTS

PART I: CANCER AS EXCESSIVE GROWTH (CANCER PARADIGM 1.0)

PART II: CANCER AS A GENETIC DISEASE (CANCER PARADIGM 2.0)

PART III: TRANSFORMATION (CANCER PARADIGM 3.0)

PART IV: PROGRESSION (CANCER PARADIGM 3.0)

PART V: METASTASIS (CANCER PARADIGM 3.0)

PART VI: TREATMENT IMPLICATIONS

PART I

CANCER AS EXCESSIVE GROWTH

(Cancer Paradigm 1.0)

1

TRENCH WARFARE

I ONCE ATTENDED a hospital meeting where the director of a new program presented its past year's accomplishments. Over a million dollars had been raised from the community for this new program, and hopes were high. I wasn't among those in the room impressed with the results being touted, but I kept quiet—because it wasn't really my business and because my mother taught me that if you don't have anything nice to say, you shouldn't say anything at all. Yet that didn't stop me from thinking that this program had wasted precious time and resources.

All around me, other participants were expressing their support. *Great job! Congratulations! Excellent work!* Even though it was obvious to everyone that there was little of value to show for the last year, most of the medical professionals around me played along with the sentiment that everything was great, just great. Nobody, myself included, stood up and yelled, "The emperor has no clothes!"

This problem is not unique to my hospital, but is pervasive in all of public health; it is how any bureaucracy functions. While keeping critical opinions to yourself is generally useful in personal relationships, it's not useful when it comes to the advancement of science. In order to solve problems, we need to know that they exist. Only then can we understand how current solutions fall short and improve on them. Lives depend on it, after all. But in medical research, opinions that dissent from the specified narrative are not welcome. This problem cuts across entire disciplines, such as the study of obesity, type 2 diabetes, and yes, cancer.

OBESITY

We are witnessing the greatest epidemic of obesity in the history of the world. Look at any statistic about global obesity and you'll find the news is bleak. In 1985, not a single American state had a prevalence of obesity above 10 percent. In 2016, the Centers for Disease Control and Prevention (CDC) reported that no state had a prevalence of obesity under 20 percent, and only three states had rates below 25 percent.[1] Yikes! We can't simply blame bad genetics, because this change has taken place within the last thirty-one years: a single generation. Clearly, we need interventions, sustainable solutions to help people lose pounds and then maintain a healthy weight.

For decades, we've fooled ourselves into believing we have a prescription for obesity: counting calories. The CDC suggests that "To lose weight, you must use up more calories than you take in. Since one pound of body fat contains approximately 3,500 calories, you need to reduce your caloric intake by 500–1000 calories per day to lose about 1–2 pounds per week." This is fairly standard advice that you can find repeated the world over by doctors and dieticians, and reported in magazines, textbooks, and newspapers. It's the same dietary advice I learned in medical school. Any physician who suggests that there is a way to lose weight by any other means is largely considered a quack. But the medical community's obsessive focus on calories has not translated into any success against the obesity epidemic. If we cannot acknowledge that our solutions fall far, far short, we will be powerless to fight the rising tide of obesity.

Few can admit that the advice to "eat less, move more" doesn't work. Yet the crucial first step toward solving the obesity epidemic is to admit to our shortcomings. Advice to count calories is neither useful nor effective. Instead, as I've argued, we must acknowledge that obesity is a hormonal imbalance rather than a caloric one. Let's embrace the truth and move forward so that we can develop interventions that actually work. Only then do we stand a chance of turning the tide on this public health crisis. As the brilliant economist John Maynard Keynes is quoted as having said, "The difficulty lies not so much in developing new ideas as in escaping from old ones."

TYPE 2 DIABETES

The horrifying epidemic of type 2 diabetes closely mirrors that of obesity. According to the CDC, about one in ten Americans suffers from type 2 diabetes. Worse, this number has risen steadily over the past few decades, with no salvation in sight (see Figure 1.1).

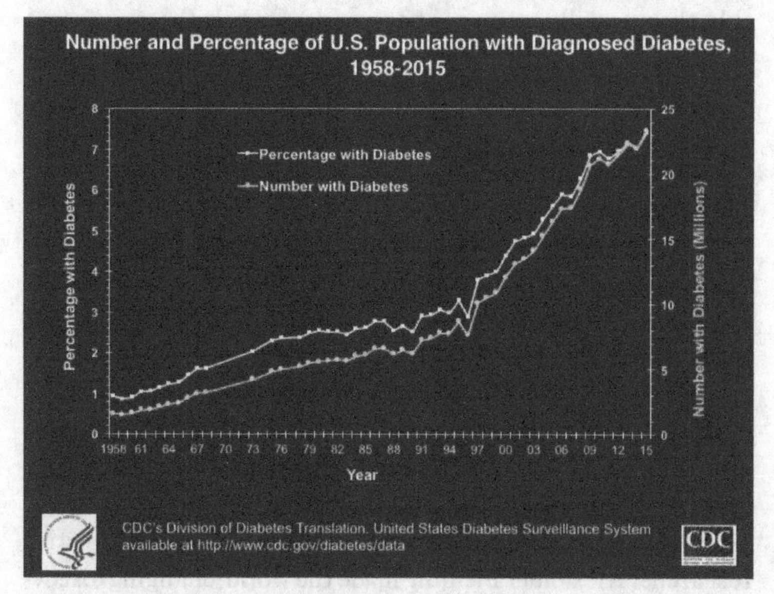

Figure 1.1

Medications that lower blood glucose, like insulin, are the standard treatment for type 2 diabetes. Over time, patients usually require higher and higher doses of these medications. If you're taking more insulin, then it's pretty obvious that your type 2 diabetes has become more severe. Yet we in the medical community (researchers, doctors) simply maintain the position that type 2 diabetes is a chronic and progressive disease, and that's just the way things are.

None of this is true. When a patient loses weight, their type 2 diabetes almost always improves. We don't need to prescribe more medication to diabetics; we need to fix their diets. But we have been unwilling to admit that our treatment approach is flawed. That would mean deviating from the agreed-upon narrative that our researchers and doctors are making brave progress against a terrible disease. Admit a problem? No way. The result? A continuing epidemic. Again, as with obesity, if we cannot acknowledge that the prevailing treatment protocol falls far, far short of acceptable, then we will continue to be powerless to help those suffering.

CANCER

This brings us, finally, to cancer. Certainly, we must be making great progress against cancer, right? Almost every day, we hear reports of some cancer breakthrough or medical miracle discovered by our pioneering scientists. Unfortunately, a sober look at the available data indicates that progress in cancer research has lagged behind that of almost every other field of medicine.

In the early twentieth century, cancer didn't attract much attention. The biggest threats to public health were infectious diseases like pneumonia, gastrointestinal infections, and tuberculosis. But public sanitation improved, and in 1928, British researcher Alexander Fleming made the world-changing discovery of penicillin. Americans' life expectancy began to climb, and the focus shifted to chronic diseases such as heart disease and cancer.

In the 1940s, the American Society for the Control of Cancer (the ASCC, which would later become the American Cancer Society) stressed the importance of early detection and aggressive treatment. The ASCC championed the routine use of the Pap smear, a gynecological screening test for cervical cancer. The results were a stunning success: with much earlier detection, death rates from cervical cancer dropped dramatically. This was an auspicious start, but death rates from other types of cancer continued to increase.

Deciding that enough was enough, then president of the United States Richard Nixon declared war on cancer in his 1971 State of the Union address, proposing "an intensive campaign to find a cure for cancer." He signed the National Cancer Act into law and injected nearly $1.6 billion into cancer research. Optimism was running high. America had ushered in the atomic age with the Manhattan Project. The country had just put a man on the moon with the Apollo program. Cancer? Surely that could be conquered, too. Some scientists enthusiastically predicted that cancer would be cured in time to celebrate America's bicentennial in 1976.

The bicentennial came and went, but the cure for cancer was nowhere closer to becoming a reality. By 1981, the tenth anniversary of the "war on cancer," the *New York Times* questioned whether this highly publicized, decade-long war had "brought real progress against this dreaded disease, or . . . been an extravagant $7.5 billion misfire?"[2] Cancer deaths continued their ruthless climb; the past decade's efforts hadn't even slowed its ascent. The war on cancer, so far, had been a complete rout.

This was not news to insiders like the National Cancer Institute's (NCI) Dr. John Bailar III, who also served as a consultant to the *New England Journal of Medicine* and a lecturer at Harvard's School of Public Health. In 1986, Dr. Bailar questioned the effectiveness of the entire cancer research program in an editorial in the *New England Journal of Medicine.*[3] In the article, Dr. Bailar noted that from 1962 to 1982, the number of Americans who died of cancer increased by 56 percent (see Figure 1.2). Adjusting for population growth, this still represented a *25 percent increase* in the rate of death from cancer, at a time when death rates from virtually every other disease were dropping quickly; crude death rates from causes other than cancer had decreased by 24 percent. Dr. Bailar noted that the data "provide no evidence that some 35 years of intense and growing efforts to improve the treatment of cancer have had much overall effect on the most fundamental measure of clinical outcome—death. Indeed, with respect to cancer as a whole, we have slowly lost ground." He wondered aloud, "Why is cancer the only major cause of death for which age-adjusted mortality rates are still increasing?"

As an insider on the cancer wars published in the most prominent medical journal in the world, Dr. Bailar had effectively yelled, "The emperor has no clothes!" He recognized the need to energize new thinking in the stultifying morass of cancer research, which had been mummifying in reiterations of the same cancer para-

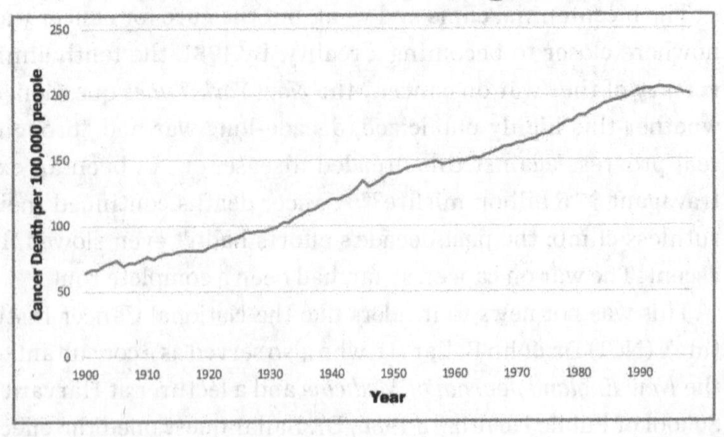

Figure 1.2: Cancer deaths, 1900–2000.

digms that had failed so utterly. In recognizing the failures of the medical community, Dr. Bailar bravely took the first step to making progress in the war on cancer.

Unfortunately, the rest of the cancer establishment was not yet ready to admit to a problem. Dr. Bailar's article received heavy criticism; it was called "erroneous" at best and "reprehensible" at worst. In the polite world of academia, this language was tantamount to the highest profanity.[4] Dr. Bailar became almost universally reviled within the community he had once led. His motives and intelligence were routinely questioned.

Vincent DeVita Jr., then the director of the NCI, called Dr. Bailar's editorial irresponsible and misleading while implying that Dr. Bailar himself had "departed with reality."[5] The president of the American Society of Clinical Oncology called Dr. Bailar "the great naysayer of our time." Ad hominem attacks were plentiful, but there was simply no denying the statistics. Cancer was getting worse, but nobody wanted to acknowledge it. The research community responded to the message by killing the messenger. *Everything is awesome*, they said, even as the bodies piled up.

Little had changed eleven years later, when Dr. Bailar published a follow-up paper titled "Cancer Undefeated."[6] The death rate from cancer had increased by *another* 2.7 percent from 1982 to 1994. The war on cancer had resulted in not just a rout, but a massacre. Still, the cancer world could not admit there was a problem. Yes, there were some notable successes. Cancer death rates for children had dropped by about 50 percent since the 1970s. But cancer is the quintessential disease of aging, so this was a major victory in a minor skirmish. Of the 529,904 deaths due to cancer in 1993, only 1,699 (3 percent) were in children. Cancer was delivering punishing uppercuts to our face, and we had managed only to tousle its fancy hairdo.

The war on cancer was reinvigorated by the continuing revelations from the study of genetics that took place throughout the 1980s and '90s. *Aha*, we thought, *cancer is a genetic disease.* A new front opened in the war on cancer, focusing our efforts on finding cancer's genetic weaknesses. A massive, multimillion-dollar international collaboration oversaw the 2003 completion of the Human Genome Project. The research community felt certain that this genetic map offered a winning battle plan against cancer. We now had a complete diagram of the entire human genome, but surprisingly, this did little to move us closer to beating cancer. In 2005, an even more ambitious program, the Cancer Genome Atlas (TCGA), was launched. Hundreds upon hundreds of human genomes were mapped in an attempt to uncover cancer's weakness. This massive research effort, too, came and went while cancer continued its progress undisturbed, calm as bathwater.

We brought our human ingenuity, massive research budgets, and fund-raising efforts to create new weapons to penetrate cancer's imperturbable shell. We believed that the war on cancer would be a high-tech battle of smart weapons. Instead, it more closely resembled the trench warfare of World War I. The front lines never moved, the war dragged on without noticeable progress, and the bodies piled up.

The stalemate in cancer stands in stark contrast to the dizzying progress in other areas of medicine. From 1969 to 2014, total deaths in the United States from heart disease dropped approxi-

Annual Mortality Heart Disease versus Cancer

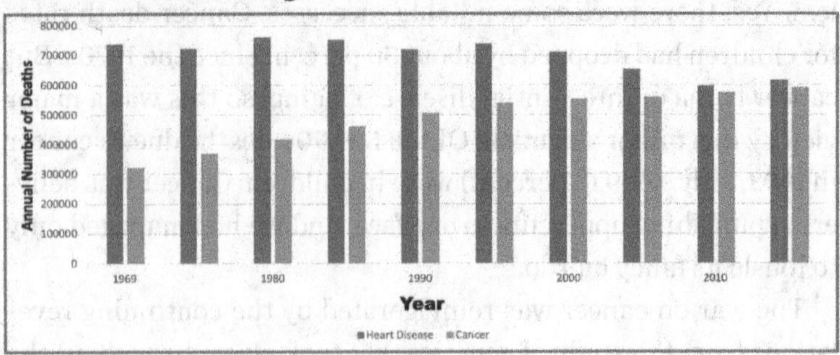

Figure 1.3

mately 17 percent despite the increasing population. But cancer? During that same time period, deaths from cancer rose a chilling *84 percent* (see Figure 1.3).

In 2009, the *New York Times* ran a headline that reflected this reality: "Advances Elusive in the Drive to Cure Cancer,"[7] noting that the adjusted death rate for cancer had dropped only 5 percent from 1950 to 2005, compared to heart disease deaths, which had dropped 64 percent, and to flu and pneumonia, which had dropped by 58 percent. Once again, an American president, this time Barack Obama, promised to "launch a new effort to conquer a disease that has touched the life of nearly every American, including me, by seeking a cure for cancer in our time."[8] Nobel Prize laureate James Watson, the co-discoverer of DNA's double helix, ruefully noted in a 2009 opinion piece published in the *New York Times* that cancer killed 560,000 Americans in 2006, 200,000 more than in 1970, the year before the "war" began.[9]

The war on cancer has not stagnated for lack of funding. The 2019 budget for the National Cancer Institute was $5.74 billion, all derived from taxpayer dollars.[10] Nonprofit organizations have proliferated like mushrooms after a rainstorm. By one count, there are more nonprofits dedicated to cancer than those for heart disease, AIDS, Alzheimer's disease, and stroke combined. The American Cancer Society generates over $800 million per year in donations to fund "the cause."

Perhaps at this point you're thinking, *But what about all the cancer breakthroughs we keep hearing about in the news? All that funding must be saving lives?* It's true that advances in treatment have been made, and these treatments have certainly made a difference. However, they're not saving as many lives as you might think.

Cancer drugs are approved by the Food and Drug Administration (FDA) if they show efficacy with minimal toxicity. But efficacy can be defined in many different ways—not all of which include saving lives. Unfortunately, from 1990 to 2002,[11] fully 68 percent of the FDA approvals were for cancer drugs that did not necessarily show an improvement in life expectancy. If these drugs didn't improve survival, what *did* they do? The most common reason for approval is called the "partial tumor response rate," which means the drugs were shown to shrink the primary tumor in volume by over 50 percent. That sounds pretty good, except when you consider that this measurement is almost completely irrelevant to survival.

Cancer is deadly because of its propensity to spread, or metastasize. Cancer is deadly because it moves, not because it is big. Cancers that don't metastasize are called "benign" because they very rarely cause significant disease. Cancers that *do* metastasize are called "malignant" because of their tendency to kill.

For example, the very common lipoma, affecting approximately 2 percent of fifty-year-olds, is a benign cancer of fat cells. It may grow to weigh up to fifty pounds. Yet despite this enormous bulk, this benign cancer is still not life-threatening. A malignant melanoma (a type of skin cancer), however, may weigh only 0.1 pound and be thousands of times deadlier because of its predisposition to spread. Once unchained, many cancers become unstoppable.

For this reason, local cancer treatments such as surgery or radiation are of limited efficacy once a cancer has metastasized. Surgeons go to great lengths in the quest to "get it all." They will cut huge swaths of normal tissue out of cancer patients to remove even the faintest rumor of a whiff of cancer cells. Surgery for cancer is performed to prevent metastasis, not because the cancer is too big. A cancer medication's ability to shrink a tumor is immaterial to overall patient survival. A drug that destroys half the tumor is no better than surgery to remove half the cancer—in other words, almost completely useless. Getting half the cancer is no better than getting none of it.

Yet the majority of new cancer drugs were approved based solely upon this questionable marker of "efficacy." From 1990 to 2002, 71 new drug approvals were granted for 45 new drugs. Of those, only 12 medications were proven to save lives, and most extended life by only a few weeks or months. In that same time, the phrase "cancer breakthrough" appeared in 691 published articles. The strange math goes like this: 691 breakthroughs = 71 cancer drug approvals = 45 new drugs = 12 drugs that barely extended patients' lives.

All these shiny new weapons in the war on cancer amounted to a jeweled handle on a broken sword. By the mid-2000s, hope for the war on cancer was fading quickly. Then a strange thing happened. We started winning.

A NEW DAWN

Amid all the doom and gloom, hopeful signs emerged. Cancer deaths, adjusted for age and population growth, peaked in the early 1990s and have now been steadily declining. What changed? Some of the credit must be given to the smoking-cessation efforts that have been consistently championed by public health officials since the 1960s. But our paradigm of understanding cancer has been slowly undergoing a revolution, and this has contributed to new treatments, which drive our recent and, hopefully, continuing progress.

The most pressing question in cancer research is the most elusive: what *is* cancer? In this decades-long war, we simply didn't know our ancient enemy. The Manhattan Project had a clear goal: split the atom. World War II had a clear foe: Adolf Hitler. The Apollo project had a concrete task: put a man on the moon and bring him back, with a little luck, alive. But what was cancer? It was a nebulous adversary, with hundreds of different variations to discern. Wars on murky ideas, like the wars on poverty, drugs, and terrorism, generally end in frustration.

Approaching a problem from the wrong angle gives you no chance of solving it. If you're not facing the right direction, it doesn't matter how fast you run; you'll never reach your destination. This book is an exploration of the story of cancer. It is not meant to offer a cure for cancer. That, for now, is still largely impossible. Instead, my goal here is to chronicle the surprising journey in our understanding of the greatest mystery of human disease. It is perhaps the strangest and most interesting story in science. What is cancer? How did it develop?

Over the last one hundred years, our understanding of cancer has undergone three major paradigm shifts. First, we considered cancer a disease of excessive growth. That's certainly true, but this did not explain why cancer was growing. Next, we considered cancer a disease of accumulated genetic mutations that caused excessive growth. Also, certainly true, but this did not explain why these genetic mutations were accumulating. Most recently, a completely new understanding of cancer has emerged.

Cancer is, improbably, a disease unlike any other we've ever faced. It is not an infection. It is not an autoimmune disease. It is not a vascular disease. It is not a disease of toxins. Cancer is originally derived from our own cells but develops into an alien species. From this paradigm of understanding, new drugs have been developed that threaten, for the first time, to end this war in the trenches.

2

THE HISTORY OF CANCER

CANCER IS A prehistoric disease, recognized since the time of the ancient Egyptians. The Edwin Smith Papyrus, translated in 1930, contained the medical teachings of the Egyptian physician Imhotep, who lived around 2625 BC. It describes a case of a "bulging mass in the breast" that was cool and hard to the touch.

Infections and abscesses are typically inflamed, and warm and painful when touched. By contrast, this mass was firm, cool, and not painful—something much worse. As for suggested treatment, the author had none. The Greek historian Herodotus, writing around 440 BC, describes Atossa, the queen of Persia, who likely suffered from inflammatory breast cancer. In a thousand-year-old grave site in Peru, mummified remains show a bone tumor, preserved by the dry desert climate. A two-million-year-old human jawbone unearthed by archeologist Louis Leakey showed evidence of lymphoma, an unusual cancer of the blood.[1] Cancer dates back at least to the dawn of humanity.

Cancer has walked this earth at least as long as we have as an ever-present adversary. Its longevity makes it unique among diseases. Maladies have come and gone. Smallpox and the Black Death once devastated the world, but have largely disappeared from the modern pantheon of health concerns. But cancer? Cancer was there in the beginning. It was there in the middle. It's still here now, worse than ever.

Despite several thousand years of advancing medical knowledge, cancer still ravages us. Cancer was likely rare in ancient times because it is a disease of aging, and life expectancy was low. If people are dying young from famine, pestilence, and war, then cancer is not a big concern.

The Greek physician Hippocrates (ca. 460 BC–ca. 370 BC), who is often called the father of modern medicine, may have appropriately named our ancient foe using the word *karkinos*, meaning "crab." This is a surprisingly astute and accurate description of cancer. Examined microscopically, cancer extends multiple spicules (a spikelike tendril) out of the main body to grab tenaciously on to adjacent tissue. Like miniature versions of its namesake, cancer distinguishes itself from other deadly diseases by its ability to scuttle around the body from one location to another. A cut on your thigh does not metastasize into a cut on your head, but a cancer in your lung can easily become a cancer in your liver.

In the second century AD, the Greek physician Galen used the term *oncos*, meaning "swelling," to describe cancer, as it was often detected as a hard nodule. From this root, the words *oncology* (the science of cancer), *oncologist* (cancer specialist), and *oncologic* (related to cancer) are all derived. Galen also used the suffix *-oma* to denote a cancer. Thus, a hepatoma is a cancer in the liver. A sarcoma is a cancer of the soft tissues. A melanoma is a cancer of the melanin-containing skin cells. Celsus (ca. 25 BC–ca. AD 50), a Roman encyclopedist who wrote the medical text *De Medicina*, translated the Greek term *karkinos* into the English word *cancer*. The word *tumor* is used to describe any localized growth of abnormal cells, which can be benign or malignant.

Cancer was first understood as an exuberant, unregulated, and uncontrolled growth of tissue. Normal tissues have well-defined growth patterns. A normal kidney, for example, grows from birth until adulthood and then stops. It then simply maintains its size, unless other diseases intervene. A normal kidney does not continue growing throughout life until it is so big that it takes up the entire abdominal space. Cancer cells, however, will continue growing until they die or you do.

Cancers are usually divided into benign and malignant varieties. Benign cancers grow, but they don't metastasize. Some examples are lipomas and basal cell carcinomas of the skin. These may become huge, but we aren't overly concerned about benign cancers, because they are rarely deadly. It is the ability to move and spread, or metastasize, that is responsible for the majority of cancer deaths.

Malignant cancer is what we usually think of as cancer, and in this book, we consider only malignant cancers. The many types of cancers (breast, colorectal, prostate, lung, myeloma, etc.) are generally named for their cell of origin. There are likely as many types of cancers as types of cells in the body. These cancers both continue growing without limit and have the ability to leave the site of origin to reestablish at a distant site.

All cancers are derived from normal cells. Breast cancer originates from normal breast cells. Prostate cancer originates from normal prostate cells. Skin cancer originates from normal skin cells. This is the particularly vexing and unusual part of cancer—that it originally derives from ourselves. Cancer is not a foreign invader. It's an internal uprising. The war on cancer is a war on ourselves.

While all types of cancer are different, this book attempts to discuss the origins of cancer as a whole, looking at the similarities among cancers rather than the differences. This is the fundamental question of this book: What turns normal cells into cancer cells in some people in some situations, but not others? In other words, what causes cancer?

The ancient Greeks believed in the humoral theory of disease, which posited that all diseases resulted from an imbalance in the four humors: blood, phlegm, yellow bile, and black bile. Inflammation was the result of too much blood; pustules, from too much phlegm; jaundice, from too much yellow bile. Cancer was considered an internal excess of black bile. Local accumulations of black bile manifested as tumors that could be palpated as nodules. However, the disease itself was a systemic excess involving the whole body.

Treatment of cancer was therefore aimed at removing this excess black bile, and included those oldies but goodies: bloodletting, purging, and laxatives. Local excision of the tumor would not work, because cancer was understood to be a systemic disease. This was yet another surprisingly astute observation from ancient physicians, and it spared many a cancer patient from surgery, which was a pretty gruesome thing in ancient times. In the absence of antiseptics, anesthetics, and analgesics, you were more likely to die of surgery than of cancer.

The humoral theory of cancer endured for centuries, but there was a big problem with it. Three of the four humors were identified—blood, lymph, and yellow bile—but where was the black bile? Doctors looked and looked, but could not find any black bile. Tumors, believed to be local outcroppings of black bile, were examined, but there was none to be found. If black bile caused cancer, where was it?

By the 1700s, lymph theory had replaced humoral theory. Cancer was believed to be caused by fermentation and degeneration of stagnant lymph that did not properly circulate. Once again, while the theory was incorrect, it contained some surprisingly astute observations about the nature of cancer. First, it recognized that cancer cells are derived from the body's own normal cells that have somehow become perverted. Second, it recognized cancer's natural tendency to spread along lymphatic drainage routes and lymph nodes.

The development of microscopes and reliable dyes to stain tissue samples allowed another major scientific leap forward. By 1838, the focus had shifted to cells rather than fluids, with the blastema theory. The German pathologist Johannes Müller showed that cancer was not caused by lymph, but instead originated from cells. Cancer, he believed, derived from the budding elements, or "blastema," between these cells. That same year, the pathologist Robert Carswell, examining several widespread cancers, was among the first to suggest that cancer may move through the bloodstream.

Cancers were simply cells, albeit bizarre-looking cells with unregulated growth. This is what I call cancer paradigm 1.0, the first great modern paradigm of understanding cancer. It is a disease of excessive growth. If the problem is too much growth, then the obvious solution is to kill it. This logic gave us surgery, radiation, and chemotherapy, and is still the basis of many of our cancer treatment protocols today.

SURGERY

Surgical treatment of cancer dates back to the second century AD, when Leonidas of Alexandria described a logical, stepwise surgery for breast cancer by removing all cancerous tissue and a margin of healthy tissue. Even with cautery to stanch the expected bleeding, surgery was fraught with danger. Surgical instruments were not sterilized. If you developed a postsurgical infection, there were no antibiotics. Most of us would not let these ancient surgeons cut our hair, let alone our bodies. One particularly macabre invention from 1653 was the breast guillotine, for amputation of the affected breast.

The advent of modern anesthesia and antisepsis transformed surgery from a barbaric, ritualistic sacrifice to a fairly reasonable medical procedure. The ancient Greeks treated cancer as a systemic disease, but nineteenth-century physicians increasingly viewed cancer as a local disease, amenable to surgery. The obvious solution, then, was simply to cut it all out—and they did. As surgical technology and knowledge grew, local tumor excision became an option in almost all cases. Whether such a procedure was useful was a different matter entirely.

Cancer inevitably recurred, usually at the incision site. Again, cancer is like a crab, sending invisible microscopic pincers out into adjoining tissue. These minuscule remnants of cancer inevitably lead to a relapse. And so, physicians began to subscribe to a new theory: if a little bit surgery is good, then perhaps a lot of surgery is even better.

In the early 1900s, Dr. William Halsted championed increasingly radical surgeries to purge breast cancer "root and stem." The word *radical*, as in a "radical mastectomy" or "radical prostatectomy," is derived from the Latin word meaning "root." In addition to the affected breast, Halsted removed a wide margin of normal tissue, including almost the entire chest wall, the pectoral muscles, and associated lymph nodes that might possibly contain the seeds of cancer. The complications were horrific, but thought to be worth it. A radical mastectomy might be disfiguring and painful, but the alternative, if cancer recurred, was death. Less invasive surgery, Halsted believed, was a misguided kindness. This became the standard surgical treatment of breast cancer for the next fifty years, making the breast guillotine look almost humane by comparison.

Halsted's results were both very good and very bad. Patients with localized cancer fared extremely well. Patients with metastatic cancer fared extremely poorly. After cancers had metastasized, the extent of the surgery was largely irrelevant, because it was a local treatment for a systemic disease. By 1948, researchers showed that less invasive surgery achieved similar local control of disease as Halsted's method, with a fraction of the surgical complications.

By the 1970s, preoperative X-rays and CT scans allowed earlier detection of metastasis, which prevented unnecessary surgery. In addition, the location of the tumor could be ascertained and the extent of surgical invasion necessary could now be precisely delineated before doctors whipped out the scalpel. Today we know that such targeted surgery is potentially curative—if the cancer is caught early. Modern technological advances have steadily reduced operative complications, and surgical deaths have dropped by over 90 percent[2] since the 1970s. Surgery remains an important weapon against cancer, but only at the proper time and in the proper situation.

RADIATION

In 1895, German physicist Wilhelm Röntgen identified X-rays, high-energy forms of electromagnetic radiation, a discovery for which he would receive the 1901 Nobel Prize. These invisible X-rays could damage and kill living tissue. Barely one year later, an American medical student, Emil Grubbe, pioneered the specialty of radiation oncology by irradiating a patient with advanced breast cancer.[3] Grubbe, also a manufacturer of vacuum tubes, had exposed his own hand to this new X-ray technology, causing an inflammatory rash, which he showed to a senior physician. Noting the tissue damage, the physician suggested that these newfangled X-rays may have other therapeutic uses, suggesting lupus or cancer as likely candidates. Fortuitously, Grubbe was caring for a patient suffering with both lupus and breast cancer at that very moment. On January 29, 1896, he exposed the breast cancer to the X-ray source for one hour. *One hour!* Modern X-ray treatments take seconds. Recalling the damage to his own hand, Grubbe did thoughtfully protect the areas around the breast cancer with the lead sheet lining from a nearby Chinese tea chest. One shudders to think of what could have happened if he had not been a tea drinker.

Meanwhile, that same year in France, physicist Henri Becquerel, along with the legendary scientists Marie and Pierre Curie, discovered spontaneous emission of radiation; the three would share a Nobel Prize for their work. In 1901, while carrying a tube of pure radium (yikes!) in his waistcoat pocket, Becquerel noted a severe burn on his skin underneath the tube. Researchers at the Hôpital Saint-Louis in Paris used his radium to develop more powerful and precise X-ray treatments. By 1903, researchers claimed to have cured a case of cervical cancer through radium treatment.[4] In 1913, the "hot-cathode tube" was used to control the quality and quantity of radiation, allowing dosing for the first time, instead of the haphazard blasting of X-rays willy-nilly against a suspected lesion.

The early period of radiation oncology, from 1900 to 1920, was dominated by the efficient Germans, who favored treatment with a few large, caustic doses of radiation. There were some impressive remissions and some impressive side effects, but few lasting cures. Burns and damage to the body were inevitable, and by 1927, French scientists realized that a single huge dose of radiation harmed the overlying skin without much affecting the cancer underneath. Instead, a smaller dose of radiation delivered over multiple days (called fractionated radiotherapy) could hit the buried target without so much surface collateral damage. This is because cancer cells are more sensitive to damage from X-rays than the surrounding normal tissue.

Fractionated radiotherapy exploits this difference in sensitivity to preferentially kill cancer cells while only injuring normal cells, which have a chance to recover. This is still the preferred method of radiation therapy today. By the 1970s, President Nixon's war on cancer supplied much-needed funds for the development of this high-tech modality.

But the biggest problem with both surgery and radiation is that they are inherently local treatments. If a cancer remained localized, then these treatments were effective, but if a cancer had metastasized, these local treatments offered little hope of recovery. Luckily, development of a more systemic treatment using chemicals (drugs) had continued concurrently.

CHEMOTHERAPY

The logical solution for a widespread cancer was to deliver "chemotherapy," a systemic, selective toxin, to destroy cancer cells wherever they hid but leave normal cells relatively unscathed. In 1935, the Office of Cancer Investigations, which would later be merged into the National Cancer Institute, set up a methodical program for cancer drug screening involving more than three thousand chemical compounds. Only two made it to clinical trials, and both eventually failed due to excessive toxicity. Finding a selective toxin was no easy task.

A breakthrough came from an unlikely source: the deadly poisonous gases used in World War I. Nitrogen mustard gas, named for its faint peppery smell, was first used in 1917 by Germany. Developed by Fritz Haber, the brilliant chemist and winner of the 1918 Nobel Prize, this deadly gas is absorbed through the skin, blistering and burning the lungs. Victims died slowly, taking up to six weeks to complete the deadly journey.

Interestingly, mustard gas has a peculiar predilection for only destroying certain parts of the bone marrow and the white blood cells.[5] In other words, it is a selective poison. In 1929, an Israeli researcher named Isaac Berenblum, studying the carcinogenic effect of tar, applied mustard gas in an attempt to provoke cancer by adding its irritant effects—but paradoxically, cancer regressed.[6]

Two doctors at Yale University hypothesized that this selective poison could be used therapeutically to kill the abnormal white blood cells in a cancer known as non-Hodgkin's lymphoma. After successful animal trials, they tested their theory on a human volunteer, now known only by his initials, J.D. This forty-eight-year-old man suffered from advanced, radiation-resistant lymphoma, with tumors in his jaw and chest so large that he couldn't swallow or cross his arms. With no other options, he agreed to the secret experimental treatment.

In August 1942, J.D. received the first dose of the mustard gas, then known only as "substance X."[7] By day four, he began to show signs of improvement. By day ten, the cancer had all but disappeared.[8] The recovery was almost miraculous, but one month later, the lymphoma relapsed, and J.D.'s medical record on December 1, 1942, contained one entry: "Died." Nevertheless, it was a great start, proving that the concept could be effective. The treatment known as chemotherapy had just been born, although war restrictions meant that the results were not published until 1946. Derivatives of mustard gas, such as chlorambucil and cyclophosphamide, are still in use today as chemotherapy drugs.

Another form of chemotherapy took advantage of folic acid metabolism. Folic acid is one of the essential B vitamins and is required for new cell production. When the body is deficient in it, new cells cannot be produced, which affects fast-growing cells like cancers. By 1948, Sidney Farber, a pathologist at the Harvard Medical School, pioneered the use of folic acid–blocking drugs in the treatment of certain types of childhood leukemia.[9] The remissions were spectacular, with cancer simply melting away. Alas, it would always come back.

The development of chemotherapy pressed onward. The 1950s witnessed some notable successes against some rare cancers. Dr. Min Chiu Li, a researcher with the National Cancer Institute, reported in 1958 that a regimen of chemotherapies had cured several cases of choriocarcinoma, a tumor of the placenta.[10] Few scientists believed him, and he was asked to leave his position at the NCI when he persisted in using his "crazy" newfangled treatments. He returned to Memorial Sloan-Kettering Hospital in New York, where his insights into chemotherapy would later be vindicated for choriocarcinoma and also metastatic testicular cancer.

The development of multiple types of chemotherapy drugs allowed more options. If one poison was not enough, why not combine multiple poisons into a chemical cocktail that no cancer cell could withstand? By the mid-1960s, Drs. Emil Freirich and Emil Frei were applying their combination of four drugs to children with leukemia, eventually increasing the remission rate to a then-unheard-of 60 percent.[11] The remission rate for advanced Hodgkin's disease rocketed from nearly zero to almost 80 percent.[12] By 1970, Hodgkin's lymphoma was considered a largely curable disease. Things were looking up. Chemotherapy had made the respectability leap from "poison" to "drug treatment."

Most chemotherapy drugs are selective poisons, preferentially killing fast-growing cells. Because cancer cells are fast growing, they are particularly susceptible to chemotherapy. If you were lucky, you could kill the cancer before you killed the patient. Fast-growing normal cells, like hair follicles and the lining of the stomach and intestines, also sustained collateral damage, leading to the well-known side effects of baldness and nausea/vomiting. Newer medications, such as many of the targeted antibodies, are not often referred to as "chemotherapy" because of the negative connotations associated with the classic medications.

CANCER PARADIGM 1.0

The first great paradigm of cancer, what I call cancer paradigm 1.0, deems cancer an unregulated growth of cells. If the problem is too much growth, then the solution is to kill. In order to kill, you need weapons of cellular mass destruction for cutting (surgery), burning (radiation), and poisoning (chemotherapy). For localized cancer, you could use locally destructive methods (surgery or radiation). For metastatic cancers, you need systemic poisons (chemotherapy).

Cancer paradigm 1.0 was a huge medical advance, but it did not answer the most fundamental questions: What was causing this uncontrolled cell growth? What was the root cause of cancer? To understand this, we need to know: what is cancer?

3

WHAT IS CANCER?

THE LEGENDARY BIOLOGIST Charles Darwin is credited as the first scientist to discuss what is called the "lumper-splitter problem."[1] In the early nineteenth century, classification was a fundamental part of natural science research. Biologists circled the globe searching for new animal and plant specimens. After careful observation, these specimens were grouped into scientific categories such as species, family, phylum, and kingdom.

In creating categories, the lumpers and splitters were opposing factions. Should certain animals be lumped together into a single category, or should they be split into separate ones? For example, humans, bears, and whales can be lumped together as mammals, but they can also be split by whether they live on land or in water. Lumping reduces the number of categories, and splitting increases them. Both contribute different but important information. Whereas splitting highlights individual differences, lumping highlights similarities.

The term *cancer* does not refer to a single disease, but denotes a collection of many different diseases related by certain qualities. Depending upon the definition used, we can identify at least one hundred different types of cancer. Traditionally, cancer biologists have been splitters, considering each cancer as a separate disease based upon its cell of origin. Cancer cells are derived from normal human cells and therefore retain many features of the original cells. For example, breast cancer cells may have hormone receptors like estrogen and progesterone, just like healthy breast cells. Prostate cancer cells produce prostate-specific antigen (PSA), just like healthy prostate cells, which can be measured in the blood.

Almost every type of cell in the human body is potentially cancerous. There are cancers of solid organs and tissues, with lung, breast, colon, prostate, and skin cancers being the most common. There are also cancers of the blood, sometimes termed "liquid" cancers, as they don't present with a single large tumor (mass of cancer cells). These include diseases such as leukemia, myeloma, and lymphoma. Each type of cell causes a different type of cancer with individual natural histories and prognoses. Breast cancer behaves and is treated completely differently from, say, acute leukemia. Splitting cancers into individual diseases can therefore be useful in treatment, but doing so highlights their differences, not their similarities. When we focus on the unique characteristics of the various types of cancer, we don't get any closer to unraveling the mystery of cancer as a single entity.

Noted cancer researchers Doug Hanahan and Robert Weinberg recognized that cancer was a collection of different diseases united by certain qualities. But what were those qualities? In all the vast literature on cancer, nobody had yet categorized the small number of principles to explain how cancers are similar. In 2000, they decided to codify the principles of malignant transformation, publishing in the journal *Cell* a seminal paper titled "The Hallmarks of Cancer."[2] The authors expected little of this, thinking their work would quickly fade into obscurity.

But something stood out about this paper, which quickly became the most influential in cancer research history. It laid the foundation for understanding cancer as a single disease rather than as many specific diseases. Hanahan and Weinberg had just become lumpers in an ocean of splitters. They asked the crucial question: what makes cancer . . . cancer?

THE HALLMARKS OF CANCER

Hanahan and Weinberg's original review in 2000 listed six hallmarks shared by most cancers. In 2011, two more hallmarks were identified and added.[3] Despite the hundreds of different types of cancer, all cancers share most of these eight commonalties, which are all critical features for survival of cancer cells. Without most of these eight hallmarks, cancer would no longer be cancer.

The Eight Hallmarks of Cancer

1. Sustaining proliferative signaling;
2. Evading growth suppressors;
3. Resisting cell death;
4. Enabling replicative immortality;
5. Inducing angiogenesis;
6. Activating invasion and metastasis;
7. Deregulating cellular energetics; and
8. Evading immune destruction.

Hallmark 1: Sustaining Proliferative Signaling

The first hallmark, and arguably the most fundamental, is that cancer cells continue to replicate and grow, whereas normal cells do not. The human body contains trillions of cells, so growth must be tightly regulated and coordinated. During childhood and adolescence, the birth of new cells outpaces the death of old cells, so the child grows larger. Upon adulthood, the number of new cells created is precisely matched by the death of old cells, so overall growth stops.

This delicate balance is lost in cancer, which continually grows, leading to abnormal collections of cancer cells, called tumors. Normal cell growth is tightly regulated by hormonal pathways, which are controlled by genes. There are genes that increase growth, called proto-oncogenes, and genes that decrease growth, called tumor suppressor genes. The two kinds of genes act like the accelerator and brakes in your car. Proto-oncogenes accelerate growth. Tumor suppressor genes decelerate growth. Normally, these genes operate in balance with each other.

Abnormal growth can occur if proto-oncogenes are excessively activated (like stepping on the gas pedal) or tumor suppressor genes are suppressed (like taking your foot off the brake pedal). In certain normal situations, such as wound healing, growth pathways are activated for a short period of time. Once the wound is healed, growth should once again slow to neutral. But cancer cells maintain this proliferative signaling, creating growth when it is no longer advantageous to do so. When genetic mutations cause excess activation of proto-oncogenes, they are then referred to as oncogenes. The first confirmed oncogene, called *src* because it caused a soft tissue cancer called a sarcoma, was discovered in the 1970s.

Cancers are not simply a giant glob of growing cells that absorb everything in their path, like the titular character in the classic science-fiction movie *The Blob*. Cancer cells face many challenges in their quest to grow into a large tumor, and even more challenges when they metastasize. At different times, a cancer must proliferate, grow new blood vessels, and break off to metastasize. A single genetic mutation is not usually capable of doing all these things, hence the need for the other hallmarks.

Hallmark 2: Evading Growth Suppressors

Many normal genes in our body actively suppress cell growth. The first tumor suppressor gene (*Rb*) was discovered in retinoblastoma, a rare type of eye cancer in children. A genetic mutation that inactivates the *Rb* gene releases the brakes on cell growth, which favors growth and hence the development of cancer.

Some of the most commonly affected genes in cancer are tumor suppressor genes, including *p53*, which is estimated to be mutated in up to 50 percent of human cancers. The well-known tumor suppressor genes called breast cancer type 1 and type 2, usually abbreviated as *BRCA1* and *BRCA2*, are estimated to be responsible for 5 percent of total breast cancer.

Hallmark 3: Resisting Cell Death

Overall tissue growth is simply the difference between how many cells are created and how many cells die. When normal cells grow old or become damaged beyond repair, they undergo programmed cell death in a process known as apoptosis. This normal cellular expiration date keeps our body running smoothly by allowing a natural turnover of cells. Red blood cells, for example, live for an average of only three months before they die, to be replaced by new red blood cells. Skin cells are replaced every few days. It's like changing the oil in your car's engine. Before putting new oil in, you must first drain the old oil out. In the body, old or damaged cells must be culled to allow room for new cells to replace them. Apoptosis is the orderly disposal of a cell when it has outlived its useful life.

Cell death occurs either through necrosis or apoptosis. Necrosis is an unintentional, uncontrolled cell death. If you accidentally hit your finger with a hammer, your cells are killed in a haphazard and disorderly fashion. The contents of a cell are splattered as when an egg hits a sidewalk. This is a huge mess, causing significant inflammation that the body must work hard to clean up. Necrosis is a toxic process that should be avoided whenever possible.

Apoptosis is an active process that requires energy. This controlled cell deletion is so crucial for survival that apoptosis has been evolutionarily conserved in living creatures ranging from fruit flies to worms to mice to humans.[4] The difference between apoptosis and necrosis is the difference between throwing a nice, well-planned dinner party and having your partner bring twenty rowdy coworkers home unannounced. Both are large dinners, but one is carefully controlled and pleasant, while the other results in a lot of chaos and yelling, with somebody eventually sleeping on the couch.

Apoptosis, this mechanism of controlled cell deletion, is common to all multicellular organisms. Allowing old cells (like skin cells) to die and replacing them with new ones rejuvenates the organism as a whole, although the individual cell must die. To avoid excessive growth, the number of old cells removed must be carefully balanced by the number of newer, replacement cells. Cancer cells resist apoptosis, changing the balance of cell division and cell death and allowing for excessive growth.[5] If fewer cells are dying, then the overall tissue is likely to grow, favoring cancer.

Hallmark 4: Enabling Replicative Immortality

In 1958, scientific dogma accepted that human cells grown in a laboratory were immortal because they could replicate themselves indefinitely. After all, a fungus or bacterium in a nutrient solution can replicate itself an infinite number of times. But Leonard Hayflick, a scientist at the Wistar Institute at the University of Pennsylvania, could not coax human cells to live past a certain life span no matter what he did. He initially worried that he was making some rudimentary mistake. Perhaps he wasn't providing the right nutrients or clearing the waste appropriately? But nothing he did could make the cells live any longer.

After three exhausting years of experimentation, he proposed the radical new idea that cells divide only a finite number of times before stopping.[6] This discovery, so fundamental for understanding both aging and cancer, was not immediately embraced by the scientific community, but took, according to Hayflick, "ten or fifteen painful years" to become generally accepted. He recalled ruefully that "To torpedo a half century old belief is not easy, even in science."[7] We now know that human cells are indeed mortal, and cannot propagate indefinitely. This limit to cellular longevity is now called the Hayflick limit.

Cells can generally replicate themselves only forty to seventy times before stopping. Hayflick correctly intuited this as a form of cellular aging, which happens in the nucleus, where the chromosomes are contained. Nobel Prize winners Elizabeth Blackburn and Carol Greider later demonstrated that cells "count" the number of replications as they progress toward the Hayflick limit with the use of telomeres, the caps on the end of the chromosomes. The telomere cap protects DNA during cell division, and each cycle shortens the telomeres. When a telomere gets too short, the cell can no longer divide, and it activates apoptosis, or programmed cell death. This process provides natural protection against the unregulated proliferation of cancer. Cellular age is not counted in years, but rather, in the number of times a cell replicates.

While normal cells are mortal, cancer cells are immortal; they, like bacteria, are not restrained by the Hayflick limit and can replicate indefinitely. Cancer cells produce an enzyme called telomerase, which increases the length of the telomeres at the end of chromosomes. Because the telomere cap never wears down, cells can keep on dividing as long as they like. This blocks both the natural cellular aging process (senescence) and timed cellular death (apoptosis). In a cell culture, you can keep growing cancer cells forever.

In what is now a well-known story, our understanding of cancer owes a tremendous debt to a woman named Henrietta Lacks. On October 4, 1951, Lacks died of cervical cancer at Johns Hopkins Hospital at the age of thirty-one. The cancerous cells removed from her body—without her consent, it should be noted—have since revolutionized medicine. For the first time, scientists propagated a cell line outside a human body indefinitely. These HeLa cells, named after Lacks, have been used in the studies of vaccines, genetics, drug development, and cancer. More than fifty million tons of HeLa cells have been grown, and they have starred in more than sixty thousand scientific papers.[8]

Normal cells, after reaching the Hayflick limit, cannot divide further. Cancer cells reproduce like digital files. You can transmit or replicate them with 100 percent fidelity to the original. From an organism's perspective, killing off defective or old cell lines keeps things running smoothly. When your clothes develop holes over time, you need to throw them out and buy new ones. That's far preferable to continuing to wear your old, faded, torn, 1970s-era bell-bottoms. When cells outlive their useful lives, they are killed and replaced. Cancer cells circumvent this apoptosis process to achieve replicative immortality.

Hallmark 5: Inducing Angiogenesis

Angiogenesis is the process of building new blood vessels, which brings in fresh supplies of oxygen and nutrients and carries away waste. As a tumor grows, new cells are situated farther from the blood vessels, just as new houses in a suburban subdivision are located farther from the main roads. New houses require the construction of new roads, and new cancer cells require the construction of new blood vessels.

Angiogenesis requires the close coordination of growth signaling of many different cell types. A breast tumor, for example, cannot simply keep making new breast cancer cells far from existing blood vessels. Somehow the cancer must induce the existing blood vessels to grow branches, just as new houses must connect their wastewater to the existing sewage system. This involves growing new smooth muscle cells, connective tissue, and endothelial cells (lining), an incredibly complex task that must be accomplished for a tumor to grow.

Hallmark 6: Activating Invasion and Metastasis

The ability to invade other tissues and metastasize is what makes cancer lethal, responsible for an estimated 90 percent of cancer deaths. Once these metastases are established, it matters little what happens to the original tumor. Cancers that cannot metastasize are called benign because they are easily treated and almost never cause death. Benign cancers share all the other five characteristics so far listed. Without the ability to metastasize, cancer is more a nuisance than a serious health concern.

Metastasis is perhaps the most difficult hallmark to achieve, requiring completion of multiple complex intervening steps. A metastatic cancer cell must first break free of its surrounding structure, where it is normally held together tightly by adhesion molecules. That's why you don't usually find breast cells floating around in the blood or the lung, for example. The freed cancer cell must survive the journey through the bloodstream and then colonize the metastatic site, a foreign environment completely different from its home. At each step along the metastatic pathway, the cancer cell acquires entirely new skill sets of incredible complexity, requiring multiple genetic mutations of existing pathways. It's like humans trying to walk on the surface of Mars without a spacesuit and expecting to flourish.

Classically, we consider metastasis to transpire late in the natural history of cancer, after a lengthy growth period of the primary tumor. We've long assumed that the cancer stayed relatively local and intact until it started shedding some cancer cells into the blood. However, newer evidence suggests that micro-metastases may be shed from the original cancer early on, but these sloughed-off cells typically do not survive.

EMERGING HALLMARKS

In 2011, Hanahan and Weinberg updated their review, adding two emerging hallmarks and two enabling characteristics, that is, traits that make it easier for cancer cells to attain their hallmarks. The first enabling characteristic is genome instability and mutations. Cancers achieve their hallmarks by mutating normal genes, and unstable genetic material makes this easier. The second enabling characteristic is tumor-promoting inflammation. An inflammatory response is a natural reaction to tissue injury or irritation. This is usually a protective response, but in some cases, it may promote cancer's progress.

Hallmark 7: Deregulating Cellular Energetics

Cells need a reliable source of energy for the hundreds of routine housekeeping tasks they undertake every day. Cellular energy is stored in a molecule called adenosine triphosphate, or ATP. There are two ways to metabolize glucose for energy: with oxygen (aerobic respiration) and without oxygen (anaerobic fermentation). A chemical process called oxidative phosphorylation, or OxPhos, is the most efficient method of energy extraction. This process burns glucose and oxygen together to generate thirty-six ATP molecules as well as a waste product, carbon dioxide, which is exhaled. OxPhos occurs in a part of the cell called the mitochondria, which are often referred to as a cell's "power plants."

When oxygen is not available, cells burn glucose using a chemical process called glycolysis, which generates only two ATP molecules, along with waste in the form of lactic acid. In the appropriate circumstance, this is a reasonable tradeoff—generating ATP far less efficiently, but without the need for oxygen. For example, high-intensity exercise such as sprinting requires large amounts of energy. Blood flow is insufficient to deliver the oxygen needed, so muscles instead use anaerobic (without oxygen) glycolysis. The lactic acid generated is responsible for the familiar muscle burn upon heavy physical exertion. This creates energy in the absence of oxygen, but generates only two ATP molecules per glucose molecule instead of thirty-six. Thus, you cannot sprint very far before your muscles tire and you must stop to rest. When blood flow becomes sufficient to clear away the lactic acid buildup, you start to recover.

For every glucose molecule, you can generate eighteen times more energy with mitochondrial OxPhos compared to glycolysis. Because of this increased efficiency, normal cells almost always use OxPhos if sufficient oxygen is available. But cancer cells, strangely, do not. Cancer cells, almost universally, use the less efficient glycolytic pathway *even in the presence of adequate oxygen.*[9] This is not a new discovery, having been first described in 1927 by Otto Warburg, one of history's greatest biochemists. This metabolic reprogramming occurs in about 80 percent of cancers and is known as the Warburg effect.

Because the Warburg effect (aerobic glycolysis) is less energy efficient, cancer requires far more glucose to sustain metabolism. To compensate, cancer cells express far more GLUT1 glucose transporters on their cell surface. This increases the rate at which cancer cells can move glucose from the blood inside the cell. The positron emission tomography (PET) scan takes advantage of the glucose avidity of cancer cells. Radioactively labeled glucose is injected into the body, and cells are given time to take it up. A scan reveals those areas that take up glucose more briskly. These "hot spots" are evidence of cancer activity.

This is a highly intriguing paradox. Cancer, which is growing rapidly, should require *more* energy, so why would it deliberately choose the *less* efficient pathway of energy generation? This is an utterly fascinating anomaly.

Hallmark 8: Evading Immune Destruction

The immune system actively seeks out and destroys cancerous cells. For example, the natural killer cells of our normal immune system constantly patrol the blood, on the lookout for foreign invaders like bacteria, viruses, and cancer cells. For this reason, patients with compromised immune systems, such as those who are HIV-positive or taking immune-suppressive drugs (like transplant recipients), are much more likely to develop cancer.

In order to survive, cancer cells must somehow evade an immune system that has been designed to kill them. While growing inside tissue, the tumor may be somewhat shielded from the immune cell that must penetrate that tissue. When cancer spreads through the blood, however, it is exposed directly and constantly surrounded by hostile immune cells.

DEFINING CANCER

The eight hallmarks represent the best scientific consensus on the characteristic behavior that demarcates what a cancer is or is not. In lumping different cancers together as a single disease, the finer details are lost, but it becomes easier to see the big picture. For example, these eight hallmarks can be further simplified into four (see Figure 3.1).

Something can be considered a cancer when it:

- Grows—it sustains proliferative signaling (hallmark 1), evades growth suppressors (2), resists cell death (3), and induces angiogenesis (5);
- Is immortal—it enables replicative immortality (4);
- Moves around—it activates invasion and metastasis (6) and evades immune destruction (8); and
- Uses the Warburg effect—it deregulates cellular energetics (7).

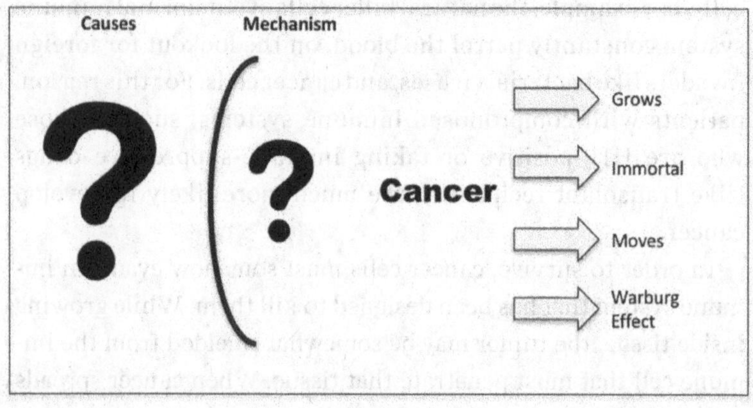

Figure 3.1

In some cases, dozens, or even hundreds, of genetic mutations are required for a cell to develop these four hallmarks. Identifying the hallmarks is a great start, but simply describing cancer's main characteristics doesn't tell us anything about why (the causes) or how (the mechanisms) cancers develop.

Many people believe that we don't know what causes cancer, but actually, we already know a great deal about it.

4

CARCINOGENS

WHAT CAUSES CANCER? That's the million-dollar (or should I say billion-dollar?) question. Most people, including many medical professionals, would respond that cancer is caused by genetic mutations. The Mayo Clinic states bluntly on its website that "Cancer is caused by changes (mutations) to the DNA within cells."[1] This is not strictly true. Except in rare cases, genetic mutations are the mechanism, not the cause, of disease. A cell that develops a number of genetic mutations becomes a cancer. That is *how* it becomes a cancer. But *why* did it develop those mutations? That is the root cause of cancer. For example, what causes lung cancer? You would be more correct to say that smoking causes cancer rather than say genetic mutations in cells X, Y, and Z cause cancer.

Factors that cause cancer to develop are called carcinogens, and we've known about them for centuries. In 1761, Dr. John Hill of London, a physician, botanist, and medical writer, described the first external cancer-causing agent, a type of smokeless tobacco.[2] Tobacco was first used by the Native Americans. The European explorers brought smallpox to the New World and brought tobacco back to the Old World. I'm not sure which agent would go on to kill more people over time. By 1614, tobacco was widely sold in Europe, with an estimated seven thousand tobacco shops in London alone. Smoking tobacco, considered rather uncouth and unsanitary, was replaced in polite society by snuffing, in which a pinch of ground-up tobacco is inhaled, or "snuffed," into the nostrils, usually after putting it on the back of the hand between the thumb and first finger. (In medicine, this area is still sometimes called the "anatomic snuff box" for that reason.) Dr. Hill described two cases of "polypusses" of the nose, which he believed were cancerous.

This was the first known description of a carcinogen, a chemical that causes cancer. Tobacco is rarely snuffed today, the practice having largely gone out of style along with use of the monocle and the topcoat, so this is of limited clinical significance.

While the association between snuffing and cancer was suggestive, more definitive proof of cancer-causing chemicals was provided by Sir Percivall Pott (1714–1788). Considered one of the greatest surgeons of his era, the London-born Pott apprenticed at St Bartholomew's Hospital and was granted the Grand Diploma by the Court of Examiners of the Company of Barbers and Surgeons.[3] After sustaining a compound ankle fracture in 1756, he brought a groundbreaking new perspective to a variety of medical topics during his forced convalescence. An astute observer of disease, he is remembered for "Pott's fracture" of the ankle; for "Pott's disease," caused by tuberculosis; and for discovering the cause of scrotal cancer.

In 1775, Pott described the burgeoning epidemic of scrotal cancer, which was the particular bane of the London chimney sweep. The Great Fire of London in 1666 had forced the legislation of new fire codes requiring smaller, more tortuous chimney configurations. This reduced the chance of another major fire, but made cleaning these new chimneys with long, straight brushes infinitely more difficult. In addition, the twisted design accumulated more soot and creosote that needed more frequent cleaning. So, chimneys were smaller, dirtier, and more difficult to clean. Solution? Send little children to clean them!

Chimney sweep apprentices started as young as three and a half, but most were older than six, only because they were otherwise considered too weak, unable to work long hours, or would too easily "go off" (die). The apprentice agreement required a weekly bath, but most followed the common London chimney sweep tradition of three baths per year. After all, why take a bath today if you're going to climb into a dusty, dirty, dangerous chimney tomorrow?

Meanwhile, in 1773, an influential Englishman named Jonas Hanway was disturbed to learn that only seven out of one hundred orphans survived over a year. Children were often assigned to workhouses, where conditions were dismal. Hanway persuaded lawmakers to limit child labor, which forced thousands of hungry children into the streets without work. For many, their only alternative to starving to death was risking their life climbing up a steaming-hot chimney to brush out some soot. Master chimney sweeps would often employ dozens of child apprentices—as many as they could afford to feed.

There were a million horrible, painful ways for London's child chimney sweeps to die. They became wedged inside chimneys, fell from great heights, suffocated when soot fell on them, or burned to death. If they survived to puberty, one more final horror often awaited them: chimney sweeps' cancer. Children as young as eight were diagnosed with cancer of the scrotum. It started with what they called a soot wart. If caught early enough, the soot wart would be cut off with a razor. But if not, the cancer would invade the skin, enter the scrotum and testicles, and then enter the abdomen. It was painfully destructive and usually fatal at that point.

This was obviously an occupational hazard, as scrotal cancer was exceedingly rare in any other circumstance. It was also quite rare outside England, where better protective clothing was available. The soot caused scrotal cancer, Pott realized, by lodging in the folds of the scrotal skin and causing chronic irritation. As the plight of the chimney sweeps was recognized, laws were passed to protect the children, and the disease faded away once again into obscurity.

The chemical benzopyrene, in coal tar, the major chemical

component of soot, was likely the main carcinogen. While soot was one of the best-studied cases of chemical carcinogens, it would be only the first of many.

ASBESTOS

In some respects, asbestos was the perfect material of the industrial age. It is an abundant natural mineral that could be woven into a lightweight fabric. It is both fireproof and a great insulator. As the world moved from the horse and buggy to steam engines, automobiles, and the great big machines of the age, the need for a fire-resistant, electricity-resistant material rose exponentially. Unfortunately, it also caused cancer.

Asbestos was an ideal compound for protective clothing, insulation, and other home products. Asbestos fibers are flexible, soft, and easy to fashion into clothing or insulation for walls and pipes. World War II provided enormous demand for fireproofing materials, particularly in navy ships. In North America, asbestos was often mixed into concrete and other construction materials to improve fire safety. It eventually found its way into North American buildings, exposing millions of people in their very own homes through insulation and heating and cooling systems.

Asbestos has been used since the time of the ancient Egyptians. Asbestos shrouds protected the embalmed bodies of pharaohs, as noted in the written records of the ancient Greek historian Herodotus. The ancient Romans wove asbestos into their tablecloths and napkins, which could then be cleaned by being simply thrown into the fire. Neat party trick.

But even then, there was an awareness of asbestos's toxic effects. The Greek geographer Strabo wrote that slave miners from asbestos quarries often suffered a "sickness of the lungs."[4] In Rome, asbestos workers tried to protect themselves by covering their nose and mouth with the thin membrane from the bladder of a goat.

Asbestos was useful and expensive, and human lives were cheap. So, whenever a flame-retardant fabric was needed, asbestos answered the call. Fireproof currency? Asbestos was used in bank notes from the 1800s by the Italian government. Fireproof clothing? The Parisian fire brigade wore jackets made of asbestos in the 1850s.

The dawning of the Industrial Revolution made asbestos a worldwide industry by the early 1900s. More than thirty million tons of it have been mined worldwide in the past one hundred years, and in that time, it has become one of the most pervasive environmental hazards.

Lung disease followed asbestos's parabolic rise in popularity. The first documented death from asbestos was recorded in 1906. Huge amounts of asbestos fibers were found in the lungs of a thirty-three-year-old asbestos textile worker at autopsy; it had effectively choked him to death from the inside. But U.S. consumption of asbestos did not peak until 1973, many decades after its health effects were known. Asbestos fibers cannot be seen, tasted, or smelled. With no immediate acute health problems, asbestos exposure can persist over many decades. The human body cannot degrade or dispose of asbestos, and once inhaled, it steadily accumulates in the lungs, causing progressive scarring.

Cancer? Yes, that was a problem, too. By 1938, reports showed that asbestos caused a rare cancer of the lining of the lung called pleural mesothelioma.[5] Recognizing asbestos as a carcinogen and admitting it were two different matters entirely, as the asbestos corporations fought tenaciously to refute the facts about their highly lucrative product.

In the 1940s, researcher Dr. Leroy Gardner proved the carcinogenic potential of asbestos when 82 percent of his experimental mice that inhaled asbestos developed cancer. This was more than a little concerning. Dr. Gardner was frantic to publicize his new scientific results, but his sponsor, the Johns-Manville Corporation, reminded him about his contractual silence. Under their research agreement, the corporation had the right of censorship. The studies, which were originally undertaken to prove the *safety* of asbestos, had instead found the opposite. But for more than four decades, none of these potentially lifesaving scientific results saw the light of day.[6]

Suppression of this vital information allowed companies to profit handsomely. Once again, as in ancient Rome, asbestos was prized while human lives were disposable. By 1973, the first lawsuit against asbestos manufacturers was won, opening the floodgates to others. This soon forced all asbestos producers into bankruptcy. Claims against asbestos manufacturers continue, constituting one of the largest mass tort actions in U.S. history. Only the intense litigation cases of the 1980s finally made public the heartbreaking correspondence between Dr. Gardner and his corporate sponsor.

In the 1950s, prior to widespread adoption of asbestos in home building materials, the estimated baseline rate of mesothelioma was 1 to 2 cases per 1 million.[7] In 1976, the incidence had jumped to 15,000 per 1 million. That's a horrifying 1.5 million percent increase in incidence.[8] Men born in the 1940s had a predicted 1 percent lifetime risk of developing mesothelioma. Deadly. From an extremely rare, almost unheard-of disease to a cancer affecting a large proportion of the entire population, mesothelioma could only be attributed to an environmental cause: asbestos. The World Health Organization (WHO) did not publish its first asbestos warnings until 1986, long after the dangers were evident.[9] Now that the horses had long ago left the barn, it was time to close the door.

Asbestos and tobacco were among the first-known chemical carcinogens, but they would not be the last. The International

Agency for Research on Cancer (IARC), part of the WHO, maintains a list of known and suspected human carcinogens classified into the following groups:

- Group 1: carcinogenic to humans;
- Group 2A: probably carcinogenic to humans;
- Group 2B: possibly carcinogenic to humans;
- Group 3: unclassifiable;
- Group 4: probably not carcinogenic to humans;

The carcinogens of group 1 include a wide range of man-made chemicals, from acetaldehyde and arsenic to vinyl chloride. But many natural substances appear on that list as well, like aflatoxin (found in moldy mushrooms) and wood dust. Certain medications are carcinogens, like the chemotherapy drug cyclophosphamide. Interesting—a drug used to cure cancer can also cause it. Radiation, sometimes used to cure cancer, can also cause it. Ironic.

As of 2018, there were 120 agents listed as group 1 carcinogens.[10] This compares to one lonely agent in group 4 (caprolactam, used to make nylon, fiber, and plastics). Strange. There are many things that can definitely cause cancer, and only a single thing that probably does *not*. (We will return to this thought later.)

RADIATION

One of the greatest scientists in the field of X-rays and radioactivity was also one of the first to die from it. Marie Curie (1867–1934) was born in Poland, the youngest of five children and a child prodigy. In 1891, she moved to Paris and met her husband, Pierre, with whom she worked in a remarkable partnership until death parted them.

In February 1898, the Curies were working with uranium-containing pitchblende and found that it was emitting far more radiation than they'd expected. Deducing the presence of a yet-unknown radioactive substance, the Curies discovered a new element they named polonium, in honor of Marie's homeland. Polonium was 330 times more radioactive than uranium.

But the remaining pitchblende was still radioactive even after extraction of the polonium, so the Curies processed the remaining material to extract the tiny quantities of yet another novel element. Only a few months after the discovery of polonium, in 1898, the Curies isolated pure radium. The notebook where Pierre Curie scrawled the word *radium*, coined from the Latin word for "ray," is still highly radioactive. Radium was the most radioactive substance that had ever been discovered.

Marie Curie was awarded the 1903 Nobel Prize in Physics for the discovery of radioactivity. Her husband, Pierre, died suddenly in 1906, in a Paris street accident, but this did not stop her prodigious scientific accomplishments. In 1911, she was awarded the Nobel Prize in Chemistry, becoming the only person in history to have won in both Physics and Chemistry.

The newly discovered element radium glowed in the dark, which quickly caught the public's fancy. Soon, radium-laced consumer products, such as glow-in-the-dark wristwatches, were being manufactured. Several million wristwatch dials were laboriously hand-painted with radium by thousands of young women. Due to the detail involved, the "Radium Girls" would moisten the brushes with their mouths, inadvertently ingesting radium-laced paint.

By 1922, it had become clear that something was very wrong, as the Radium Girls began literally disintegrating. Their teeth fell out for seemingly no reason. One dentist noted that as he prodded gently, an entire jawbone broke apart. By 1923, this severe bone deterioration was so well known that it was called "radium jaw." The ingested radium had lodged in the bones of the jaw and continually emitted radiation that had essentially burned away the bone and surrounding tissue. One Radium Girl died when the tissues of her throat degenerated and she hemorrhaged out her jugular vein. Another, walking around her pitch-black home, noted that her bones were glowing in the mirror. Her body had absorbed so much radium that she was literally glowing, a "ghost girl." Those whose bodies were not crumbling to dust often grew huge, monstrous-size cancers of the soft tissue, called sarcomas. By the 1930s, it had become a well-established fact that chronic radiation exposure caused cancer.

Today, radiation workers routinely wear protective lead gowns, but Marie Curie and her colleagues worked day in and day out completely unprotected, in an environment bombarded with the most potent radiation known. They, too, were not spared from the horrors of radiation illness as, one by one, they died mysteriously. Marie Currie's decades of radiation exposure left her chronically sick, as the radium destroyed her bone marrow (aplastic anemia). In 1995, when Marie and Pierre Curie's bodies were transferred to the Panthéon in Paris, so the two could be honored among France's most important historical figures, lead-lined coffins were used to protect visitors from their dangerously radioactive remains. They will stay in these protective cases for at least another *fifteen hundred* years. Marie Curie's personal notes and artifacts, on public display, are also highly radioactive.

Marie and Pierre Curie's daughter, Irène Joliot-Curie, and son-in-law, Frédéric Joliot-Curie, took up the baton and continued the family's pioneering work on radiation. The two co-discovered artificial radioactivity, for which they were awarded the Nobel Prize in Chemistry in 1935. But Irène would also not be spared the curse of radiation sickness. She died of leukemia at age fifty-seven at the Curie Institute hospital in Paris.

The risk of cancer increases linearly with the radiation dose. Radiation is classified as ionizing or non-ionizing. Ionizing radiation carries enough energy to disrupt molecular bonds to break them apart into ions, damaging a cell's DNA, and the surviving cells are left with unstable chromosomes more prone to mutations when the cell replicates.[11] Radiation has been listed as a group 1 carcinogen for decades now. Non-ionizing radiation is less intense, so it can often be dissipated without lasting tissue damage.

While chronic radiation is a carcinogen, acute radiation may not be as carcinogenic as initially feared. At the tail end of World War II, the American bomber *Enola Gay* dropped the first atomic bomb on the Japanese city of Hiroshima on August 6, 1945. Its fiery kiss killed an estimated eighty thousand people instantly, and more died later due to radiation exposure and burns.[12] But the biggest worry for the survivors was the latent risk of cancer due to this massive radiation exposure. In 1950 the Atomic Bomb Casualty Commission (ABCC) and the Life Span Study (LSS) monitored both atomic bomb survivors and their children over the next sixty-five years. While there was certainly an excess of cancers, the magnitude was not nearly as dire as had been feared. The following figure shows the excess rates of cancers that can be

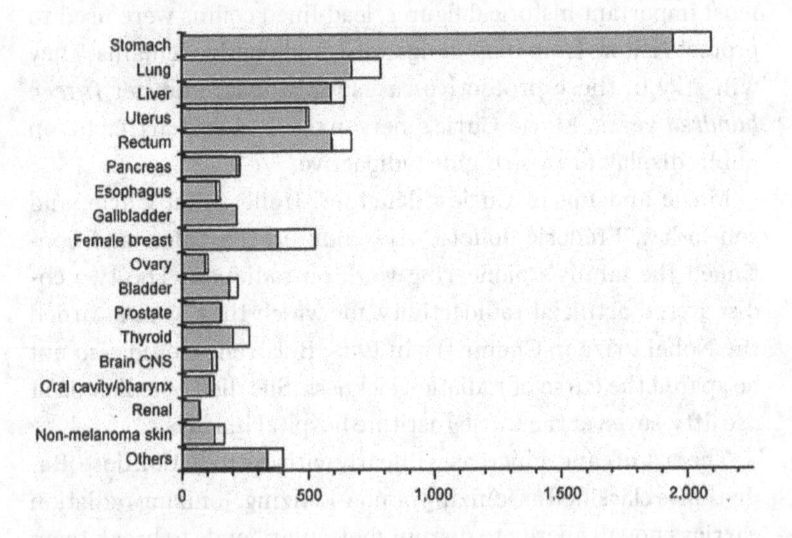

B. R. Jordan, "The Hiroshima/Nagasaki Survivor Studies: Discrepancies between Results and General Perception," *Genetics* 203, no. 4 (2016): 1505–12.

Figure 4.1

attributed to the atomic bomb in the clear bars.[13] The shaded areas show the baseline risk of cancer (see Figure 4.1).

Atomic bomb survivors are generally perceived as being affected heavily by cancer, with horribly deformed children, but the reality is, thankfully, quite different. Cancer rates did increase, but minimally (usually by less than 5 percent), and life expectancy was shortened by only months. The risk was real, but the magnitude was largely imagined.

All life on earth is constantly exposed to naturally occurring ionizing radiation that emanates from outer space. Cells protect themselves with increased antioxidant defenses and radiation-induced apoptosis.[14] When cells are irreversibly damaged by radiation, they undergo a ritualistic suicide and are removed from the body. So, let's return to our initial question: what causes cancer (see Figure 4.2)?

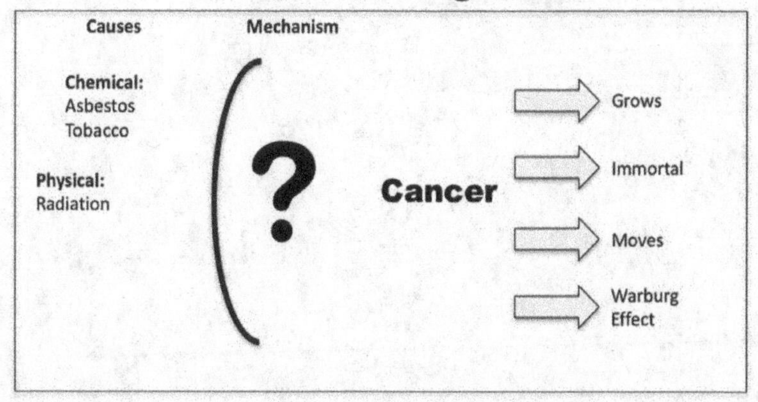

Figure 4.2

We knew that certain chemical agents caused cancer. We also knew that physical agents such as radiation caused cancer (see Figure 4.2). But soon, an outlandish theory was proposed: what if cancer could be caused by a virus?

5

CANCER GOES VIRAL

THE IRISH SURGEON Denis Parsons Burkitt was only eleven years old when an injury destroyed his vision in one eye. He threw himself into his studies, and after completing surgical training, he joined the Irish Army Medical Corps. He was stationed in Africa, where he would make his most important discoveries. What he lost in eyesight he made up for in insight, becoming one of the most influential doctors of his era.

In 1957, Burkitt was astounded to treat a five-year-old boy with multiple jaw tumors. In all his years of medical training, he had never seen such a thing. But this was only the first of many such patients with strange tumors. Shortly afterward, he saw a second child with four tumors in his jaw and multiple tumors in his abdomen. The biopsy showed "small round cell sarcoma." It was cancer.

Two children presenting with this highly unusual (for him) cancer in short succession piqued Burkitt's curiosity. Reviewing the local hospital records, he found a staggering twenty-nine other children with similar cancers. This type of cancer was apparently common in Africa, but Burkitt knew nothing of it, and there was no mention of it anywhere in the medical literature. In 1958, he published his findings in the *British Journal of Surgery*.[1]

Not all of Africa was similarly afflicted. Shortly after the publication of his paper, local cancer specialists showed Burkitt that this particular cancer, common in certain parts of Africa, was simply not seen in South Africa. Intrigued, Burkitt began tracing Africa's "lymphoma belt, which ran through the middle of the African continent" (see Figure 5.1).[2] Sure enough, cancer followed a definite geographical distribution. In his mapping, he determined that elevation above sea level and distance from the equator were major factors in the incidence of cancer. This suggested that temperature was a key determinant of a population's susceptibility to the disease. In Africa, this type of regional distribution of illness was not unusual. Infectious diseases spread by mosquitos, for example, followed an identical pattern. But this was a cancer, not an infection.

A cancer-causing virus? Perhaps the notion was not as silly as it first seemed. In 1910, Peyton Rous at the Rockefeller Institute, a chicken virologist, transmitted a sarcoma from one chicken to

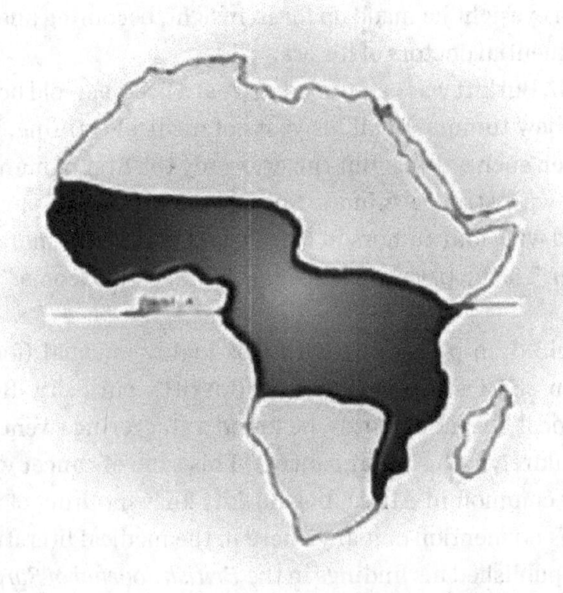

I. Magrath, "Denis Burkitt and the African Lymphoma," *Ecancermedicalscience* 3, no. 159 (2009): doi: 10.3332/ecancer.2009.159.

Figure 5.1

another. The cancer-causing agent was named the Rous sarcoma virus (RSV), for which Rous received the 1966 Nobel Prize in Medicine. In 1935, a papillomavirus was found to cause tumors in rabbits, and in the 1940s, leukemia-causing viruses were isolated in mice and cats. But could a virus cause cancer in humans? Viral cancer might be true for a few chickens in a few research laboratories at the frontiers of medicine, but this was practically unknown in clinical medicine. But the data is the data, and it does not much care what you think.

The African children's cancer was common in areas where the temperature did not fall below sixty degrees Fahrenheit and where they had at least twenty inches of rainfall per year—precisely the conditions required for mosquito breeding. Africa's lymphoma belt was essentially identical to the endemic areas for malaria, yellow fever, and trypanosomiasis (sleeping sickness), all conditions spread by mosquitos. Burkitt suspected that this cancer, now renamed Burkitt's lymphoma, was linked to an infection. In 1961, Burkitt sent some tumor samples to London to be examined by pathologist Michael Anthony Epstein, who had access to modern electron microscopy.

Laboriously growing the tumor cells in culture, Epstein identified a herpes-like virus particle.[3] This previously unknown virus, and the first known human cancer-causing virus, is now called the Epstein-Barr virus (EBV). It turns out to be one of the most common viruses in the world, with an estimated 90 percent of adults having been exposed to it.[4] In developed countries, initial EBV infection usually occurs during adolescence, sometimes accompanied by symptoms of infectious mononucleosis, or "mono." Transmitted by saliva, EBV is sometimes called the "kissing disease." In Africa, however, initial infection usually occurs at birth. In Uganda, for example, an estimated 80 percent of children under age one have been exposed to EBV, compared to less than 50 percent in America. If almost the entire world is infected with EBV, why did only some children get cancer? And why was lymphoma confined to the "lymphoma belt"? These are good questions for which firm answers still do not exist.

Burkitt's lymphoma might be caused by the coinfection of EBV with malaria.[5] In the 1960s, the African islands of Zanzibar and Pemba sprayed the toxic insecticide DDT to eradicate mosquitos. Malaria rates dramatically plunged, from 70 percent to 5 percent, and lymphoma rates followed this steep decline. When the highly toxic DDT was banned, malarial rates gradually rose, along with lymphoma, as inseparable as salt from the ocean.

In Tanzania, malaria prophylaxis with chloroquine reduced Burkitt's lymphoma by an astounding 82 percent. When increasing drug resistance forced the discontinuation of this program, malarial rates rose, and lymphoma rates increased 273 percent.[6] The precise mechanism of disease is still uncertain, but malaria may stimulate overproduction of B lymphocytes (the malignant cells in lymphoma). These cells then become infected with EBV, which somehow triggers their transformation into cancerous cells.

In other parts of the world, EBV causes a completely different cancer, called nasopharyngeal cancer (NPC). It is a rare cancer worldwide, but common in Hong Kong, Taiwan, and in the native Inuit people of Alaska and Greenland. In 2012, it accounted for only 0.71 percent of cancers worldwide, but 71 percent of those cases occurred in Southeast Asia.[7] It is unknown why EBV should cause different disease in different populations despite almost universal exposure. In southern China, it is the third most common cancer[8] and more than ten times more common than in Europe and America.

As with Burkitt's lymphoma, NPC is associated with EBV infection in early childhood. In Hong Kong, almost 100 percent of children have been exposed to EBV by age ten. Asian immigrants to other countries suffer far less NPC, a fact that argues against a purely genetic predisposition. The risk of NPC falls by about 50 percent in Chinese immigrants to America.[9] Some have speculated that a once-popular diet staple in China, salted fish, may be the missing link. The salt preservation process was inefficient in China, allowing for significant putrefaction and the development of the chemical N-nitrosamine, a known carcinogen.

THE SPECIAL VIRUS CANCER PROGRAM

This discovery, that cancer could be caused by infections, was electrifying. It opened up the terrifying possibility that cancer was contagious. Yet, like the hope remaining in Pandora's box, it also offered the possibility that it was curable. Bacteria could be killed with antibiotics. While antiviral medications had not yet been developed, vaccines were available, and once instituted widely, these became extremely effective at eradicating viruses and preventing viral outbreaks. Viral infections like measles, mumps, polio, and chicken pox, once a rite of passage in childhood, have largely faded.

The National Cancer Institute raced to investigate the exciting new possibilities. In 1964, the Special Virus Cancer Program (SVCP) was launched with the mandate of identifying additional viral causes of cancer. Over the next decade, the SVCP received over 10 percent of the total research funds earmarked for cancer—almost $500 million. By contrast, the research funds devoted to investigating the role of diet in cancer amounted to less than one twentieth that sum.

The SVCP was an enormous undertaking, and it was the beating heart of President Nixon's war on cancer. Hundreds of monkeys were inoculated with human tumors to see if these could be transmitted. Yet, in the end, the project produced little useful data. The SVCP itself was held in rather low esteem by the scientific community and was regarded as having political rather than scientific goals.[10] Scientists suspected that its true purpose was to give the appearance of making progress rather than making real progress. One prominent researcher noted that "The SVCP has been extremely ineffective and maybe has even had a negative effect." Other researchers cynically opined that its unwritten motto should have been "Nothing too stupid to test."

Lack of oversight led to contractors awarding themselves multimillion-dollar deals, and the *New York Times* reported that the managers of the SVCP "are also often the recipients of large amounts of money they dispense." In 1974, due to sharp criticism by the National Cancer Advisory Board about conflicts of interests, the SVCP was reorganized.[11] When one looks back at the project now, *boondoggle* is the word that comes to mind.

The program was formally terminated in 1980, with most of the scientific community convinced by the fiasco that infections and viruses had little or nothing to do with cancer. Then, only a few years later, new evidence surfaced that, once again, pointed to infections as the key cause of certain cancers.

HEPATITIS B AND C

Viral hepatitis (inflammation of the liver) has been described in the medical literature for millennia. The most conspicuous manifestation is jaundice, a noticeable yellowing of the eyes and skin. The first virus identified, hepatitis A, is typically found in overcrowded cities and army barracks and is transmitted through fecal contamination. Hepatitis A causes an acute illness but not chronic disease. Other forms of infectious hepatitis that caused chronic liver disease were transmitted through contamination of body fluids, such as blood and sexual contact.

Through the early twentieth century, the increased use of syringes inadvertently increased the spread of viral hepatitis. Syringes and needles were expensive, so they were routinely reused, often without adequate sterilization. In 1885, an outbreak of jaundice followed a mass vaccination of shipyard workers in Bremen, Germany, and also occurred in an asylum for the mentally ill in Merzig, Germany, where 25 percent of the vaccine recipients developed jaundice. Blood transfusions, a practice that increased exponentially during World War II, were also a risk factor for viral hepatitis. By 1947, the distinction "hepatitis B" was accepted, but the virus itself had not yet been identified. Then came Dr. Barry Blumberg, who would eventually win the 1976 Nobel Prize in Medicine.

Blumberg was an American physician and geneticist whose primary research interest was the diversity of populations, not liver disease or viruses. As he studied the variety of proteins in human blood, it occurred to him that blood transfusion may cause new proteins to form. In 1961, he discovered a new protein that he called the Australia antigen, as it was discovered in the blood serum of an Australian aboriginal man.[12] Following the trail of the Australia antigen would eventually lead Blumberg to his momentous discovery of the hepatitis B virus, one of the smallest DNA viruses to afflict humans. Endemic in Asia, the hepatitis B virus is often transmitted from mother to child, resulting in many asymptomatic children becoming chronically infected, which greatly increases their chance of developing liver cancer.

In 1981, studies found that chronic hepatitis B infection increased the risk of liver cancer by two hundred times.[13] In 2008, liver cancer was the fifth most common cancer in men and the seventh most common in women, worldwide. China in particular accounts for about 50 percent of those cases and deaths.[14]

Hepatitis B vaccines became available by the early 1980s, and nationwide vaccination programs in Asia have virtually eradicated liver cancer in the pediatric population. Hepatitis B vaccination has now been incorporated into the national infant immunization programs of at least 177 countries worldwide. Chronic infections and liver disease have dramatically declined, with beneficial implications for the future of liver cancer.

After identification of hepatitis B in the 1960s, post-transfusion hepatitis decreased, but it did not disappear, which implied another yet-to-be-identified blood-borne virus that could cause chronic liver disease.[15] This became known as "non-A, non-B hepatitis" because it was . . . well, not hepatitis A and not hepatitis B. (Scientists are sometimes a funny lot. It's not clear to me why nobody immediately said, "Guys, it's not hepatitis A, and it's not hepatitis B. So, can we all agree to call it, I don't know, hepatitis C?")

The identification of the hepatitis C virus took until 1989 because the amount of virus in the blood is several thousandfold lower than in hepatitis B. Both hepatitis B and C cause chronic liver disease. At its peak, hepatitis C infected approximately 160 million people worldwide and caused liver cancer in many of those afflicted. It is predominantly transmitted by sharing infected needles. In the post–World War II era, the reusing of needles for vaccines, especially in Italy, caused early hepatitis C outbreaks. Thereafter, the main route of transmission was illicit drug users who shared needles. Today, cutting-edge antiviral drugs may cure up to 90 percent of those infected with the virus, offering significant hope for the future.

Liver cancer develops only after many decades of chronic infection and inflammation. About 80 percent of liver cancer is related to the hepatitis B (HBV) and hepatitis C (HCV) viruses. HBV is estimated to cause 50 to 55 percent of cancer and HCV, 25 to 30 percent.

HUMAN PAPILLOMAVIRUS

In the 1970s, Dr. Harald zur Hausen, at the German Cancer Research Center in Heidelberg, noted a large number of scientific reports of genital warts "converting" into cancers in women. Human papillomavirus (HPV), of which there are hundreds of different subtypes, was known to cause genital warts. Based on this observation, he reasonably proposed that HPV caused both genital warts and cervical cancer. Cancer researchers, stung by the recent fiasco of the Special Virus Cancer Program, were not particularly receptive to his theory. Later, in an interview with the Nobel Prize committee, zur Hausen recalled, "My proposal was not welcome at the time."[16]

Excited by his observation, zur Hausen focused his research on HPV. In 1979, he first isolated HPV subtype 6 from genital warts, but this subtype was not linked in any way to cervical cancer. Tenaciously, he next isolated HPV subtype 11, but this, too, was largely irrelevant to cervical cancer. In 1983, he isolated HPV subtype 16. Bingo! Viral DNA from HPV subtype 16 was found in about half of all cases of cervical cancer. Zur Hausen had just uncovered incontrovertible evidence that HPV subtype 16 infection played a major role in cervical cancer. A year later, he cloned both HPV 16 and 18, the two subtypes of HPV now known to cause the majority of cervical cancers.

By 1999, HPV was found in 99.7 percent of invasive cervical cancer.[17] There are more than a hundred types of HPV, of which thirteen are cancer causing. Types 16 and 18 are the most common in North America, accounting for 70 to 80 percent of cervical cancers. It would take zur Hausen more than a decade to amass the scientific proof that would earn him the Nobel Prize in Medicine in 2008.

The revolutionary cycle had come full circle—from identifying and isolating the virus, to detection within the cancer cell, to the development of vaccines that protect up to 95 percent against HPV. Even today, cervical cancer is a significant worldwide burden. In 2012, there were an estimated 500,000 new cases worldwide and 266,000 deaths,[18] but vaccination programs starting in 2007 against HPV subtypes 16 and 18 have already reduced infection and the risk of premalignant warts by over 50 percent.[19] The long-held dream of cancer vaccination is quickly coming true.

HELICOBACTER PYLORI

One of the most bewildering successes in the war on cancer was the startling worldwide progress against stomach cancer. What made it puzzling was that for decades, researchers had absolutely no idea why stomach cancer was retreating. It was like winning Wimbledon without even knowing how to play tennis. Stomach cancer is particularly deadly because of the lack of early warning symptoms. By the time it is diagnosed, it is often too late.

It was not a trivial success, either. In the 1930s, stomach cancer was the most common cause of cancer death in the United States and Europe.[20] Yet, by 2019, it ranked only as the seventh leading cause of cancer-related deaths in the United States. One of the most virulent cancers in the world was steadily losing momentum, but we had just about no idea why.

Rates of stomach cancer vary greatly worldwide. The Japanese suffer ten times more stomach cancer than Americans, but when they move to the United States, their risk of stomach cancer drops dramatically, a fact that points to an environmental rather than a genetic problem. The risk of stomach cancer is much higher for a Japanese person in Japan compared to a Japanese person in America. What could account for the wide variation in stomach cancer and for its steady decline? The answer came from an unlikely source: two obscure Australian physicians studying stomach ulcers.

In 1981, Drs. Barry Marshall and Robin Warren were looking at some strange-looking bacteria on pathology slides taken from the stomachs of patients. These bacteria had been observed for over a century, but they were casually dismissed as random stains created while slides were being processed. At the time, everybody believed that the stomach was a completely sterile environment. It was thought that stomach acid created a harsh, hostile, and highly acidic environment that killed all bacteria. In the 1980s, any scientist would have considered the possibility that a bacterium could survive in the stomach to be laughable—any scientist except for Marshall and Warren, that is. Convinced that these bacteria were real, Marshall tried to grow them from biopsy specimens. He failed with his first thirty-three attempts, but on patients thirty-four and thirty-five, the lab techs made a fortuitous mistake.

Bacterial cultures were routinely discarded after two days, based on the assumption that there were no viable bacteria. Accidently left in the incubator too long, the cultures of patients thirty-four and thirty-five turned positive. The bacteria were there, but took much longer than usual to grow. Marshall identified the slow-growing bacteria *Helicobacter pylori* (*H. pylori*) as the causative agent of peptic ulcer disease.

Remarkably, *H. pylori* can persist for decades in the stomach. It uses the protein urease to neutralize its highly acidic surroundings and grows in its own protective cloaking device. Genetic studies show that *H. pylori* has been colonizing human stomachs for more than 58,000 years.[21] *H. pylori* was hiding in plain sight this whole time.

Marshall had a problem, though. Nobody believed him. In desperation, he cultured the bacteria from a patient with gastritis, "swizzled the organisms around in a cloudy broth, and drank it."[22] Super gross—but effective. Five days later, Marshall developed a stomach infection, proving that *H. pylori* caused the inflammation that led to an ulcer.

This was a stunning revelation. Until the 1980s, practically every physician and researcher in the world believed that stomach ulcers were caused by too much stress. Treatment of peptic ulcer disease consisted mainly of trying to relax. As you might imagine, long walks in the woods and meditation were not particularly effective against this infection.

Understanding that most stomach ulcers were caused by bacteria meant that antibiotics were curative. A cocktail of three medications, including two different antibiotics taken for one to two weeks, now cures about 80 percent of cases of *H. pylori* infection.[23] For these insights, Marshall and Warren were awarded the 2005 Nobel Prize in Medicine.

Approximately half the world's population is infected with *H. pylori*, although the vast majority of those afflicted have no symptoms. The urban overcrowding and poor sanitation in many parts of Asia laid the groundwork for much higher rates of *H. pylori* infestation. By the mid-1990s, the striking similarity between the worldwide prevalence of *H. pylori* infection and stomach cancer was noted. In Korea, for example, a nation with one of the highest rates of stomach cancer, 90 percent of adults over age twenty have *H. pylori*.[24] It soon became clear that *H. pylori* caused not only chronic infection and ulcers but also stomach cancer.

H. pylori infection is associated with up to a sixteenfold increased risk of cancer.[25] *H. pylori* infection starts with chronic inflammation (gastritis), which progresses to atrophy, metaplasia, dysplasia, and then finally, cancer. In 1994, the International Agency for Research on Cancer (IARC) listed *H. pylori* as a group 1 carcinogen (definite) in humans. It is estimated to be responsible by itself for 5.5 percent of the global cancer burden.[26]

H. pylori infestation has receded in recent decades, thanks to improved sanitation and housing conditions. Less *H. pylori* means less stomach cancer—the likely key to our stunning success in reducing stomach cancers. We were winning the war without even knowing why. Eradication of *H. pylori* with antibiotics reduced the chronic inflammation that leads to premalignant lesions of the stomach.[27] A rarer form of stomach cancer, known as mucosa-associated lymphoid tissue (MALT lymphoma) also occurs in those infected with *H. pylori*. In its early stages, MALT lymphomas can be completely cured by eradication of *H. pylori*.[28] Of those infected with *H. pylori*, only 10 percent will develop peptic ulcer disease, 1 to 3 percent develop stomach cancer, and less than 1 percent develop MALT lymphoma.[29] But when multiplied by half the world's population, these numbers become significant.

CANCER PARADIGMS

Let's return to our original question: what causes cancer? Chemical carcinogens such as asbestos, tobacco, and soot cause cancer. Physical carcinogens such as radiation cause cancer.

Infections, both viral and bacterial, are also carcinogens—and they're not so rare, with an estimated 18 percent of cancers having infectious disease origins.[30] The main agents are *Helicobacter pylori*, human papillomaviruses, hepatitis B and C viruses, Epstein-Barr virus, HIV, and a few others.

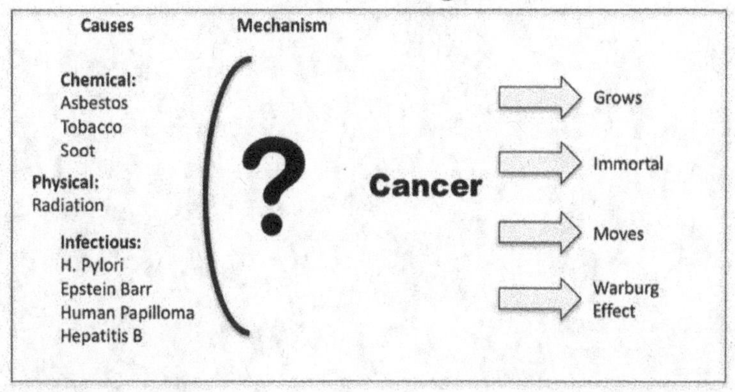

Figure 5.2

By the 1960s, all the pieces seemed to be falling into place. We knew many of the underlying factors that cause cancer. But in considering cancer as a whole, what do these diverse factors have in common? What is the unifying mechanism? To this important question, cancer paradigm 1.0 had no answer (see Figure 5.2). But by the 1970s, a new paradigm of understanding was being constructed.

PART II

CANCER AS A GENETIC DISEASE

(Cancer Paradigm 2.0)

6

THE SOMATIC MUTATION THEORY

THE GENETIC REVOLUTION

In 1866, Gregor Mendel launched the field of genetics with the publication of his now-legendary paper on plant hybridization, which described, among other things, wrinkly versus round peas. The word *genetics* was coined in 1906 by biologist William Bateson to denote the burgeoning new "science of heredity and variation."[1] Traits such as eye color and hair color are passed from generation to generation, encoded by sections of deoxyribonucleic acid (DNA) called genes, which are contained within chromosomes.

The German biologist Theodor Boveri noted in 1902 that some sea urchin eggs with an abnormal number of chromosomes grew rather exuberantly, much like cancer cells. He guessed that certain genes within the chromosomes stimulate growth and that mutations of these genes caused the excessive growth.[2] Boveri also hypothesized that other genes were responsible for stopping growth. If you cut yourself, your body must activate genes that signal cells to multiply and heal the wound. Once the wound is healed, other genes must tell the cell to stop growing. Boveri laid out his basic hypothesis in his 1914 book, *The Origin of Malignant Tumours*.[3]

Boveri's basic postulates were proven correct with the discovery of those exact genes, now called oncogenes (genes that promote cell growth) and tumor suppressor genes (genes that suppress cell growth). The first human oncogene was identified in the 1970s, when it was discovered that certain strains of the Rous sarcoma virus (RSV) caused cancer in chickens but other strains did not. By comparing the two viral genomes, researchers isolated the *src* gene responsible for the cancerous transformation, the world's first oncogene. In 1976, Nobel laureates Harold Varmus and Mike Bishop transformed cancer genetics by discovering the human equivalent of the *src* gene, immediately transforming *src* from a viral oddity in chickens to a key player in the genetics of most human (and animal) cancers.

Most cancers contain numerous changes to both oncogenes and/or tumor suppressor genes. *Src* normally increases cell growth the way the accelerator of a car increases motion. RSV causes a mutation in *src*, inappropriately activating it, leading to the unregulated growth seen in cancer. By the end of the 1970s, two other highly prevalent human oncogenes were discovered, the *myc* and the *egfr* genes.[4]

Tumor suppressor genes normally stop cellular growth like the brakes of a car stop its motion. A mutation that inactivates these genes will therefore promote cell growth, just as releasing the brakes will make a car go faster. The *p53* tumor suppressor gene, identified in 1979, is the most frequently mutated gene in human cancer.[5]

These new discoveries seemed to offer a perfect explanation for why cancer cells grow so quickly. Both activating mutations of oncogenes and inactivating mutations of tumor suppressor genes could accelerate cell growth, leading to cancer. This coalesced into the widely accepted somatic mutation theory (SMT), which views cancer primarily as a disease caused by accumulated genetic mutations. Somatic cells include all cells of the body except for germ line cells, the cells responsible for sexual reproduction, such as the sperm and egg. Mutations in these somatic cells (such as breast, lung, or prostate) accumulate, and a random aggregation of these mutations may be enough to cause cancer. This view of cancer, what I call cancer paradigm 2.0 (see Figure 6.1), dominated cancer research in the 1970s and is still championed today by the American Cancer Society, which states plainly that "Cancer is caused by changes in a cell's DNA—its genetic 'blueprint.'"[6]

Supporting this view, researchers postulated that specific inherited genetic mutations caused cancer without the need for an external agent. Familial, or inherited, cancers are relatively unusual, accounting for only approximately 5 percent of cancers, leaving the vast majority (95 percent) of cancers as sporadic mutations. Nevertheless, the SMT proved that cancer could be as simple as a disease of genetic mutations.

For example, a single inherited genetic mutation in the retinoblastoma tumor suppressor gene causes rare eye cancers in children. An inherited mutation of the von Hippel–Lindau tumor suppressor gene leads to an increased risk of kidney cancers. In breast cancer, the *BRCA1* and *2* genes are the best-known susceptibility genes that confer a high risk of breast cancer, but these account for only an estimated 5 percent of breast cancer cases. Overall, the contribution of inherited genetic defects to cancer is small, but those rare cases confirmed the underlying unifying mechanism of carcinogenesis.

Inherited mutations caused cancer. Chemicals, radiation, and viruses could also cause genetic mutations or other changes in the genetic code leading to the dysregulated growth of cancer. Bingo! The pieces of the puzzle were fitting together *perfectly*.

Rarely, a single mutation is sufficient to turn normal cells into cancer cells. A normal cell contains various mechanisms to repair damaged DNA, so if the damage is minor, it can often be rectified. But if the DNA repairs cannot keep up with the inflicted damage, then mutations accumulate. When several critical mutations converge, cancer results. Most common cancers require multiple mutations.

But how did the mutations accumulate? Asbestos or tobacco smoke or radiation can cause genetic change, but these were not targeted to any specific gene or chromosome. The implicit answer from the SMT was that these mutations were not planned, but that they accumulated more or less randomly. It is simply bad luck when all the critical mutations occur together.

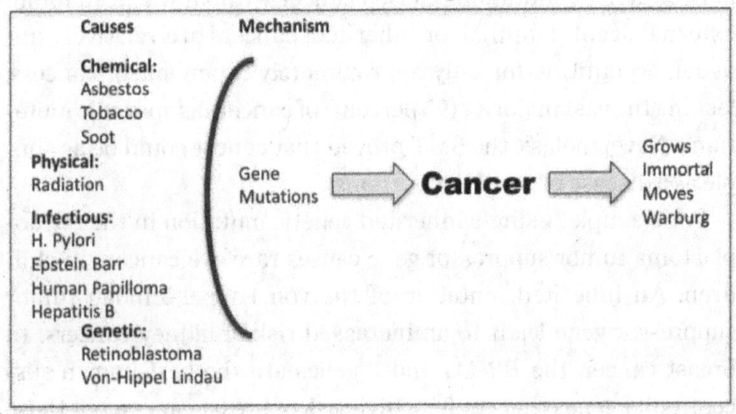

Figure 6.1

The new genetic tools developed in the 1970s showed that cancer cells were indeed chockablock full of genetic mutations. By the 1980s, animal models confirmed that chemicals, radiation, and viruses, the known causes of cancer, could mutate oncogenes and tumor suppressor genes to cause cancer. When mice were exposed to chemical carcinogens, they developed skin cancer, and those cancers had mutations in their oncogenes.[7]

Chemicals, X-rays, viruses, and inherited genetic disorders all have vastly different physiologic effects, but all caused cancer. The common thread was that they all caused DNA damage and gene mutations. A carcinogen causes cancer because it is mutagenic—that is, it increases the rate of gene mutation. Given that mutations accumulate randomly, more mutations increase the risk of cancer, just as buying more lottery tickets increases the chances of winning the jackpot.

The SMT suggested the following chain of events, represented in Figure 6.2:

1. Normal somatic cells (e.g., lung, breast, or prostate) sustain DNA damage.
2. If the rate of DNA injury exceeds the rate of repair, then random genes become mutated.
3. A chance mutation in a gene controlling growth (oncogenes or tumor suppressor genes) causes exuberant and sustained growth. This is an important first step toward cancerous transformation, but not the only one, because growth represents only one of many hallmarks of cancer.
4. Other gene mutations accumulate randomly over time. When certain critical abilities (hallmarks) coalesce, the cell fully transforms into a cancer.

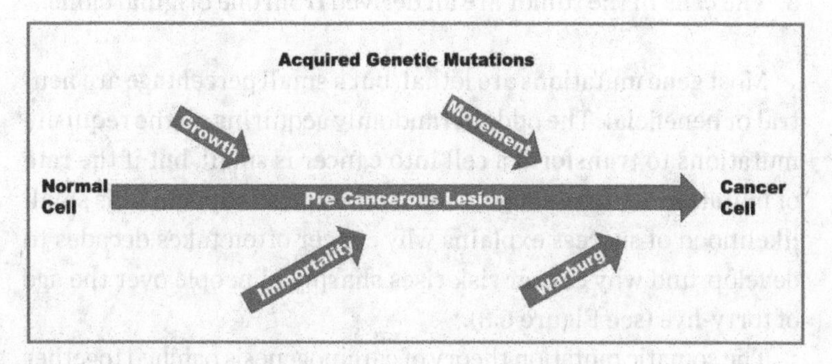

Figure 6.2

Most common cancers need multiple mutations. It is like the game of baseball. A big hit, like a home run, scores all by itself. A single horrendous mutation, such as retinoblastoma, may result in cancer. But in baseball, you can also score runs by putting together multiple hits. Multiple gene mutations can also combine to become cancer. Increasing mutation rates—say, with tobacco smoking—increases the risk of mutations. Given enough mutations, cells will eventually, by chance, become cancerous, just as an infinite number of monkeys randomly pounding the keys of an infinite number of typewriters will eventually produce the novel *War and Peace*.

These random mutations bestow all the "superpowers" needed for cancer to thrive. The ability to constantly grow, to become immortal, to move around, and to use the Warburg effect are all above and beyond what normal cells may do. Having accumulated all the superpowers that define cancer cell behavior, these cells replicate and grow. The resulting mass of cancer cells, the tumor, is a genetic clone of this original cancer cell.

The basic postulates of SMT include:

1. Cancer is caused by acquiring multiple DNA mutations.
2. These mutations accumulate randomly.
3. The cells in the tumor are all derived from one original clone.

Most gene mutations are lethal, but a small percentage are neutral or beneficial. The odds of randomly acquiring all the requisite mutations to transform a cell into cancer is small, but if the rate of mutation is high enough, then it is bound to happen. This small likelihood of success explains why cancer often takes decades to develop, and why cancer risk rises sharply in people over the age of forty-five (see Figure 6.3).[8]

The somatic mutation theory of carcinogenesis patched together all the disparate known causes of cancer into a coherent, unified theory. This paradigm focused research from extrinsic agents (chemicals, radiation, and viruses) onto intrinsic defects (genetic mutations). All these various carcinogenic insults created the seed

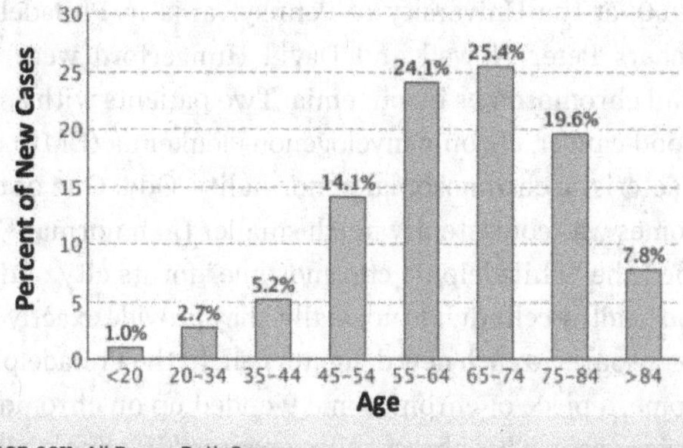

NCI, "Age and Cancer Risk," National Cancer Institute, April 29, 2015, https://www.cancer.gov/about-cancer/causes-prevention/risk/age.

Figure 6.3

of cancer by causing gene mutations. While both the seed and the soil are important for growth, it seemed that the seed, according to the SMT, was the most important component. Cancerous cells are similar to normal cells because they were derived from normal cells. Cancer cells weren't alien intruders, but mutated versions of our own cells. We had seen the enemy, and it was ourselves.

The SMT was a breakthrough, promising new directions for research and new treatments. Cancer was now viewed as a cell-centered problem of genetic mutations. If we could find and then treat those mutations, logic followed, we could cure cancer. SMT led to some astounding predictions and some astounding successes. Rather than simply using the traditional tools of cancer medicine—to cut, burn, or poison—we could use incredibly precise molecular tools to develop entirely new pharmaceutical protocols to cure cancer. By the 1980s, the SMT made good on that promise, delivering one of the most spectacular weapons yet seen in the war on cancer.

THE PHILADELPHIA CHROMOSOME

In 1960, at the University of Pennsylvania in Philadelphia, researchers Peter Nowell and David Hungerford were studying human chromosomes in leukemia. Two patients with a rare type of blood cancer, chronic myelogenous leukemia (CML), shared a characteristic chromosomal abnormality. Odd. One of the chromosomes was consistently much smaller than normal.[9] This was dubbed the "Philadelphia chromosome" for its city of discovery. When healthy cells divide normally, they provide exactly the same chromosomes to each new daughter cell. In the Philadelphia chromosome, a piece of chromosome 9 ended up on chromosome 12, and vice versa. This abnormality occurred in almost all cases of CML, and was exclusive to CML—no other types of cancer displayed this characteristic.

The Philadelphia chromosome produced an abnormal protein known as the bcr/abl kinase, a protein that toggles cell growth on or off precisely, depending upon the situation. The abnormal bcr/abl protein turned cell growth "on" and never turned it off. This uncontrolled growth eventually led to cancer. Researchers hunted for a drug to block this kinase, and in 1993, the drug firm Ciba-Geigy (now Novartis) selected the most promising candidate, called imatinib, to undergo human trials.

Human drug trials typically comprise three phases. Phase 1 studies are designed only to assess drug toxicity. This allows researchers to establish a safe dose so that further investigation may determine the drug's effectiveness. In these early trials, imatinib improved CML in an astounding fifty-three of the fifty-four patients who took over 300 mg/day. This was a miracle. Researchers would have been happy if nobody died during this phase, but instead they found a virtual cure. Better yet, there was no evidence of significant drug toxicity at this dose.

The larger, phase 2 trials test efficacy, and about two thirds of investigational drugs end their journey here. Pharmaceutical researchers are generally happy if their drug kills a few cancer cells and manages not to kill any patients. Imatinib breezed through phase 2 like an Olympic hurdler. An unheard-of 95 percent of early-stage CML patients completely cleared their leukemic cells. Even more astounding, the Philadelphia chromosome could no longer be detected in 60 percent of treated patients. This drug didn't just kill CML cancer cells; it was essentially curing the cancer.

It was a miracle drug, but more exciting, it provided a proof of concept for this new genetic paradigm of cancer. Imatinib was going to be the vanguard in the coming onslaught of new, targeted medications that promised superior efficacy with lower toxicity than standard treatments like chemotherapy. As we've discussed, chemo medications are selective poisons that kill cancer cells slightly faster than normal cells. If chemo could be considered a kind of carpet bombing, then this new generation of drugs would be the "smart bombs" of the cancer arsenal, homing in on specific targets to destroy cancer cells without causing much collateral damage.

Imatinib, known as Gleevec in the United States, is the unquestioned superstar of the genetics-focused approach to cancer treatment. Before the introduction of imatinib, CML was responsible for taking roughly 2,300 American lives per year; in 2009, after imatinib treatments began, annual CML deaths were reduced to 470. This oral medication with virtually no side effects was so dramatically successful that it was considered to herald an entirely new era of precision-targeted chemotherapy.

With the introduction of imatinib, science marked the dawn of a new age of genetic "cures" for cancer. On the cover of its May 28, 2001, issue, *Time* magazine proclaimed that "There is new ammunition in the war against cancer. These are the bullets."—alongside a picture of imatinib. It was an entirely new and better way of treating cancer, just in time for the new century.

The genetic paradigm of cancer had proven its mettle in the crucible of battle. Finding the precise genetic abnormality led to identifying the abnormal protein, which led to discovering a medication to neutralize that protein, virtually curing that particular cancer. Yes, CML was a relatively rare cancer, but this was only the beginning. Soon, another major victory would be achieved in breast cancer, with the development of the drug trastuzumab. Unlike CML, breast cancer was a major-league cancer, second only to lung cancer in causing cancer deaths in women.

HER2/NEU

In 1979, researcher Robert Weinberg at the Massachusetts Institute of Technology was chasing oncogenes. He discovered a cancer-causing segment of DNA taken from neurologic tumors in rats that he named *neu*. The human equivalent was discovered in 1987 as human epidermal growth factor receptor 2 (*HER2*), so this gene became known as *HER2/neu*, a potent oncogene. Up to 30 percent of all breast cancer cases overexpressed the *HER2/neu* gene by up to one hundred times normal. Those cancers are much more aggressive and often deadlier than those without.

The newly established, soon-to-be giant drug company Genentech located the *HER2/neu* gene using DNA probes, but the question remained: how were they going to block it? Standard drugs are small molecules that can be synthesized in a chemical plant, but none of these specifically blocked the *HER2* protein as imatinib had done so successfully with the bcr/abl kinase. But by the 1980s, the technology of the genetic revolution had advanced substantially, and Genentech pioneered an entirely new class of therapy that would offer another major leap forward in cancer treatment.

A healthy immune system produces proteins called antibodies to help fight off foreign invaders. Antibodies are very specific in their targets. For example, infection with the measles virus stimulates the body to produce antibodies that recognize measles.

After you successfully fight off the infection, your body retains these antibodies. If you are reexposed to measles, your preexisting antibodies instantly recognize the virus and activate the immune system to destroy it. This is why it is rare to develop measles infections more than once in your life. Antibodies work by recognizing specific DNA sequences, and Genentech insightfully recognized that *HER2/neu* is also just a DNA sequence.

In a remarkable feat of genetic engineering, scientists at Genentech created a mouse antibody that could bind and block the *HER2/neu* protein. But a mouse antibody injected into a human would be instantly recognized as foreign and destroyed by the human immune system. Genentech's ingenious solution was to create a mouse–human hybrid antibody to block the *HER2/neu* gene with high specificity, which became the drug called trastuzumab (Herceptin).

But there was yet another problem. Only about 30 percent of breast cancers carry the abnormal *HER2/neu* gene, so administering this very costly drug to every breast cancer patient would be extraordinarily wasteful and prohibitively expensive. So, in another innovative leap forward, scientists developed a simple test for the gene. Now only those patients with cancers overexpressing the abnormal *HER2/neu* were given trastuzumab.

This exciting development ushered in a new era in therapeutics. Drugs would not only be precision-guided weapons but also personalized. A drug did not need to work for every patient suffering from a disease in order to help a subset of those patients. We could define and treat only those expected to benefit. This approach saved money and spared patients from unnecessary side effects. Amazing. Medicine had finally found the holy grail of genetic therapy. If we could identify those few mutations that drove cancer for each specific person, we could then select the proper drug or antibody to use. Treatments could be personalized through genetic testing to reverse and potentially cure the disease.

Even before FDA approval, breast cancer patients implored

Genentech to release the drug on compassionate grounds. Nobody yet knew if it worked, but patients with metastatic breast cancer had no other options, and trastuzumab was a gleaming beacon of hope. In 1995, Genentech set up the first-ever FDA-approved expanded-access program for cancer drugs. Its hunch was correct. By 1998, trastuzumab was approved by the FDA for *HER2*-positive breast cancer and ready for prime time. By 2005, human trials showed that Herceptin cut the risk of breast cancer deaths by about one third.[10] The genetic age of precision, personalized cancer medicine had begun gloriously. It would be all rainbows and unicorns from here on, right?

CANCER PARADIGM 2.0

The genetic revolution had led us, by the early 2000s, to a major threshold. Our arsenal in the war on cancer had so far consisted of indiscriminate ways of killing cells: cutting (surgery), burning (radiation), and poisoning (chemotherapy). Blasting a cancer into oblivion seemed so crude compared to using highly specific gene-targeted antibodies to deliver a deadly payload of toxins. Only the "bad guys" were killed, sparing the collateral damage commonly seen with the older treatments. Victory seemed inevitable, as we landed body blow after body blow against cancer. We had new weapons that could penetrate cancer's tough carapace. We had new defenses against cancer's deadly pincers. The next step was to map out the one or two genetic mutations for each cancer the same way we had done for CML and the *HER2/neu*-positive subset of breast cancer.

Imatinib proved that the concept worked in "liquid" tumors of the blood, such as CML, and trastuzumab proved that this concept worked for solid tumors, too. It was only a matter of finding the mutations of various cancers and then designing the right drugs to destroy them.

Treatment - Cancer Paradigm 2.0

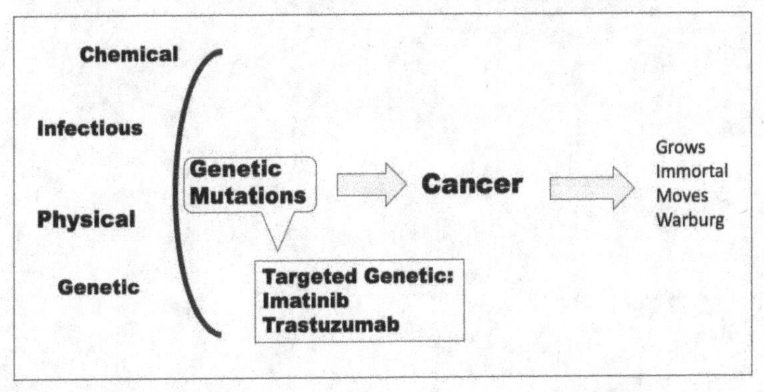

Figure 6.4

The genomic revolution was unstoppable and showed no sign of slowing. Rather, the pace of technological advance and medical knowledge was accelerating. The new drugs, while difficult to develop, were priced accordingly, and profits from the first few were prodigious. Countless start-ups, big pharmaceuticals, and universities alike joined in the new gold rush. With the map of the human genome in hand, finding the mutations that were then blocking researchers from curing cancer would be as easy as shooting fish in a barrel.

Our understanding of cancer had progressed substantially from a "disease of excessive growth" to a "disease of genetic mutations that was causing the excessive growth" (see Figure 6.4). We had peeled back a layer of the truth of cancer's origin: carcinogens cause cancer by causing gene mutations. Now that we understood the underlying root cause of cancer, we could develop lifesaving drugs.

As we rounded the corner of the twenty-first century, it seemed to many that we were on the precipice of a cancer-free world. Imatinib and trastuzumab were instant hits. But like so many other one-hit wonders, the first hit turned out to be the best.

7

CANCER'S PROCRUSTEAN BED

IN 2013, THE Academy Award–winning actress Angelina Jolie made the choice to have both her breasts surgically removed after learning she had tested positive for the *BRCA1* gene mutation, which substantially increased her risk of breast and ovarian cancer. Her mother had died of ovarian cancer at age fifty-six, and Jolie's grandmother and aunt were also cancer victims. Jolie, thirty-eight and the mother of six children, decided to undergo a preventive double mastectomy to avoid the same fate; two years later, she had her ovaries removed, forcing her body into premature menopause.

More than half a century had passed since the discovery of DNA's double helix. Where were all the promised genetic miracle cures? In the end, patients—even incredibly wealthy and influential celebrities—must surgically remove their breasts and ovaries to prevent cancer. It would seem that our genetic wizardry was making little more progress against cancer than the breast guillotine. How did things go so wrong with the genetic paradigm of cancer?

TWIN STUDIES

The clearest evidence against a predominantly genetic basis for cancer comes from twin studies. Identical twins share identical genes, whereas fraternal twins share only 50 percent genetic material on average, the same as any two siblings. Twins reared in the same household also share similar environmental influences. Comparing identical to fraternal twins gives you an idea of how much genes influence cancer rates.[1]

A large study of the twin registries of Sweden, Denmark, and Finland concluded that the majority of the risk in the causation of cancer is *not* genetic. In fact, genetics accounts only for an underwhelming 27 percent of risk. The vast majority of the risk (73 percent) is environmental. The authors concluded that "Inherited genetic factors make a minor contribution to susceptibility to most types of neoplasms." The environment plays the principal role in the development of cancer.

This statistic holds true even for those with the *BRCA1* gene, which is often referred to as a "breast cancer death sentence." The risk of developing breast cancer in patients with *BRCA1* and *2* by age fifty is 24 percent for those born before 1940, but 67 percent for those born after.

The predominant problem is not the gene itself, but the environment that allows these cancerous tendencies to manifest.[2] In other words, cancer growth depends upon not only the seed but also, and more importantly, the soil. Even in cancers with a known high genetic risk, a person's environment plays the dominant role in whether cancer develops. In most common cancers, genetics contributes only an estimated 20 to 40 percent of the risk (see Figure 7.1).

A thirty-two-year twin registry follow-up revealed that the increased risk of cancer in fraternal twins was only 5 percent; it was 14 percent in identical twins.[3] Certainly, there was a genetic link to cancer, but it was hardly the overwhelming certainty it is often made out to be. Cancer is largely caused by environmental, not genetic, factors. This becomes even clearer when we look at the changes to cancer risk when a population suddenly changes environments.

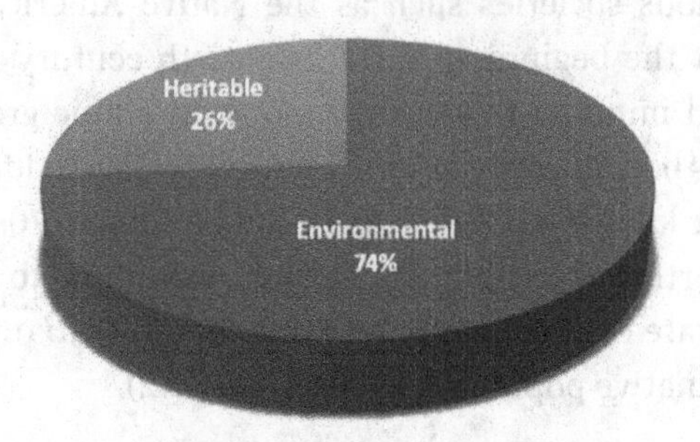

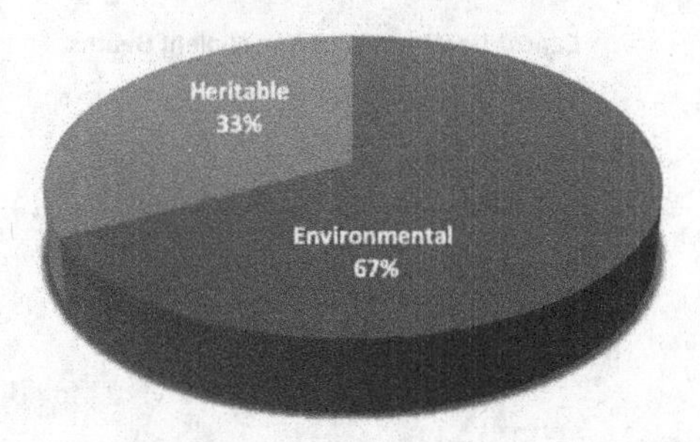

P. Lichtenstein et al., "Environmental and Heritable Factors in the Causation of Cancer," *New England Journal of Medicine* 343 (2000): 78–85.

Figure 7.1: Factors in Causation of Cancer

ABORIGINAL POPULATIONS

Cancer certainly existed in ancient societies—it has been identified in mummified remains of ancient Egyptians—but unlike today, its incidence was incredibly rare.[4] Unfortunately, analyzing fossil and mummy records is not particularly enlightening because the expected life span of ancient peoples was generally much shorter than our own, and we know cancer risk increases with age. But there are other examples of societies where lifestyles had changed in a relatively short period of time, such as the indigenous peoples of North America.

Indigenous societies such as the Native Americans and Canadians at the beginning of the twentieth century were largely considered immune to cancer. Among all ethnic groups, Native Americans had the lowest rate of cancer.[5] By the middle of the century, it was known but still rare. In the 1960s and '70s, the Ojibwa tribe in northwestern Ontario, Canada, was found to have an age-adjusted rate of cancer that was only one half to one third that of the nonnative populations (see Figure 7.2).[6]

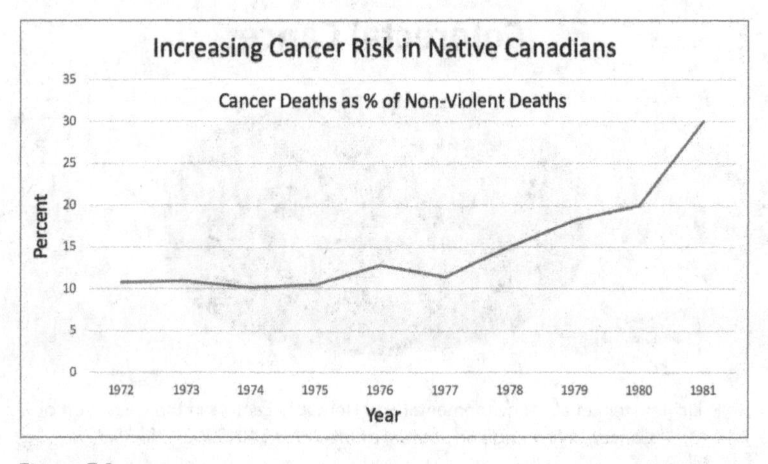

Figure 7.2

Cancer rates of the native Ojibwa population rose sharply in the 1980s, coinciding with the increasing Western influence on their lifestyle. Their gene pool would not have changed significantly over a few decades, pointing once again to the massive effect of environment, predominantly lifestyle and diet, on cancer incidence. In other words, the genetic "seed" may be the same, but the changing "soil" of the environment significantly alters the risk of cancer.

This sharp increase in cancer occurred despite a steadily decreasing rate of smoking in native populations since the early 1970s, which should have decreased the rate of cancer. From 1975 to 1981, the rate of cancer deaths as a percentage of nonviolent deaths approximately tripled, from 10 to 30 percent. If cancer were largely caused by genetic mutations, then what was causing these mutations? Diet and lifestyle factors are not mutagenic, but those factors clearly had a major influence on cancer rates—which posed a big problem with the SMT model.

The cancer experience of the Canadian Inuit of the Arctic mirrors that of other native populations. Early descriptions from 1923 suggest that cancer was essentially nonexistent among the Inuit;[7] a 1949 report uncovered only fourteen cases of cancer in a ten-year span.[8]

After World War II, however, the Inuit, forced out of their traditional lands, began living in larger urban centers, and their lifestyle changed from one of subsistence hunting to a more Westernized way of living, with service and trade as the backbone. Their traditional diet of mostly fish and sea mammals (very low in carbohydrates and vegetables and high in protein and fat) changed to include imported foods, most of which contain refined grains and sugars. As the Inuit lifestyle shifted in the 1950s, the age-adjusted rate of cancer more than doubled (see Figure 7.3).

The types of cancers afflicting the Inuit were changing, too.[9] Traditionally, the Inuit suffered from cancers caused by the Epstein-Barr virus, including cancers of the nasopharynx and salivary glands. The most common cancers in the white population then, as now, were lung, breast, and colon. While the rates of traditional, EBV-related cancers among the Inuit did not increase from 1950 to 1997, the rates of lifestyle-associated cancers, typical of the white population, rose precipitously.[10]

Because the genetic makeup of native populations was virtually unchanged, and lifestyle is not mutagenic, the somatic mutation theory failed to explain how the environmental change led to

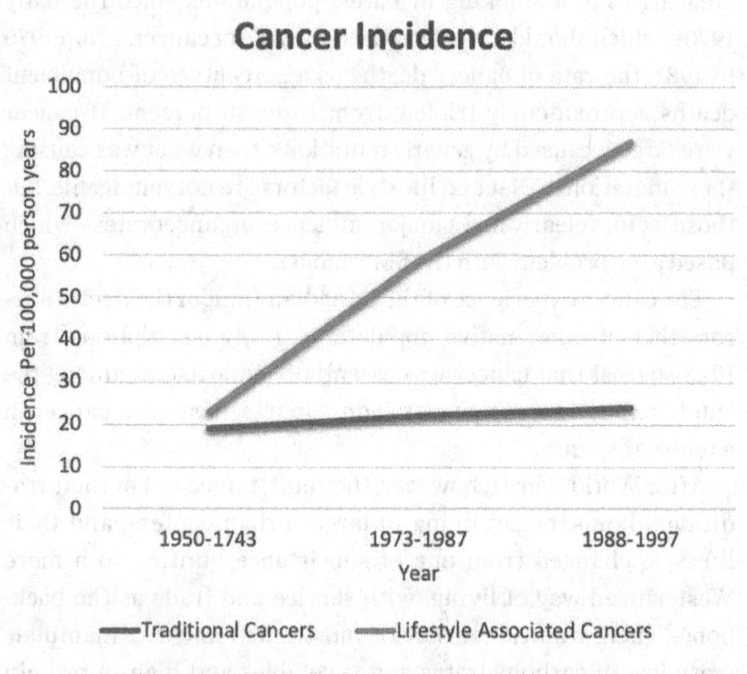

J. T. Friborg and M. Melbye, "Cancer Patterns in Inuit Populations," *Lancet Oncology* 9, no. 9 (2008): 892–900.

Figure 7.3: Cancer patterns in Inuit populations.

this dramatic increase in cancer rates. The genetic "seed" was the same, but the changing "soil" of the environment, including diet and lifestyle, altered the risk of cancer substantially.

MIGRANT STUDIES

The SMT assumed that carcinogens act by increasing the rates of genetic mutation, and it predicted that migration from one country to another should not substantially change cancer rates. A Japanese woman's moving to the United States should not substantially change her risk of breast cancer. A Japanese man's moving to the United States should not substantially change his rate of prostate cancer. But it does—and by a heck of a lot.

The breast cancer rate in the United States is two to four times higher than that of China or Japan, even for immigrants. A Chinese woman who moves to San Francisco doubles her risk of breast cancer compared to that same Chinese woman in Shanghai. Within a few generations, the cancer risk for that Chinese immigrant woman's family approximates that of a white woman in San Francisco (see Figure 7.4). The data are similar for Japanese immigrants to the United States.

The rates of cancer are highly dependent upon the environment—mainly diet and lifestyle.

Cancer rates are lowest for Asian women in Asia, intermediate for Asian immigrants to America, and highest for Asian American women born and raised in America.[11]

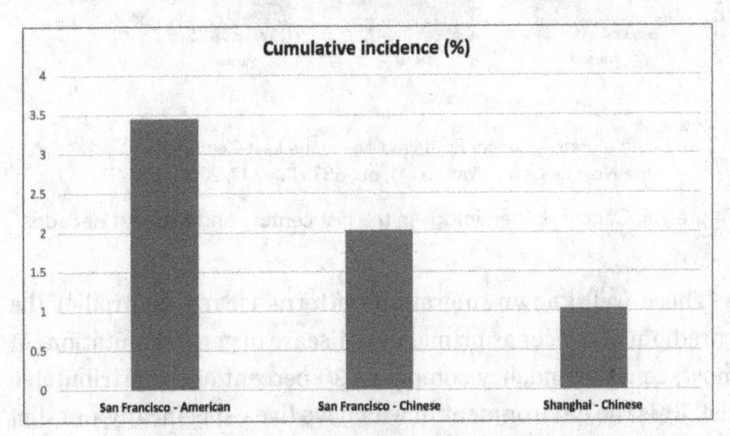

R. G. Ziegler et al., "Migration Patterns and Breast Cancer Risk in Asian-American Women," *Journal of the National Cancer Institute* 85, no. 22 (November 17, 1993): 1819–27.

Figure 7.4: Changing breast cancer risk with migration.

The same phenomenon can be observed in other cancers. A Japanese man who immigrates to Hawaii experiences about seven times the risk of prostate cancer compared to a Japanese man living in Osaka, Japan. For stomach cancer, we see the reverse; the rate of stomach cancer for a Japanese man in Japan was almost five times the rate for a Japanese immigrant to Hawaii. In this case, we know this reduced risk is likely the result of reduced exposure to bacterial infestation with *H. pylori* (see Figure 7.5).[12]

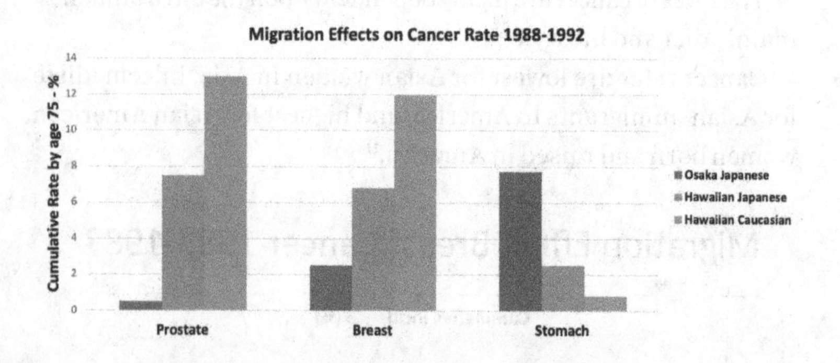

J. Peto, "Cancer Epidemiology in the Last Century and the Next Decade," *Nature* 411, no. 6835 (May 17, 2001): 390–95.

Figure 7.5: Cancer epidemiology in the last century and the next decade.

These well-known migration patterns clearly contradict the paradigm of cancer as primarily a disease of genetic mutation. At most, genetic tendency comprises 30 percent of the attributable risk. It is the environment in which we live—specifically, our diet and lifestyle—that carries the most weight in determining our risk of cancer. The genetic paradigm of cancer had focused myopically on the "seed," but it was both the seed *and* the soil *and* their interaction that determined cancer risk.

THE HUMAN GENOME PROJECT AND BEYOND

The SMT paradigm ruled cancer research from the 1970s until the 2010s like a ruthless dictator. Researchers not conforming to the dictates of the SMT quickly found themselves in the research equivalent of the Siberian gulag. The war on cancer during the reign of the SMT made glacial progress. Survival rates for the most common types of cancer barely budged. People were not living longer. Cancer rates were not dropping. The standstill in cancer research was in marked contrast to almost every other human endeavor of the late twentieth century.

Technology—from biotechnology and genetics to computers and semiconductors—was advancing at a pace never before seen in human history. Network connectivity (internet, social media) advanced at breakneck speed. Computing power was doubling every eighteen months or so. Space travel had become a reality. Within medicine, heart disease deaths were dropping rapidly as advances in medications, monitoring, intervention, and surgery combined to reduce these deaths almost by half.

But cancer? Cancer was a delinquent child. We certainly weren't ignoring it. There was no lack of money. Cancer research consumed billions of dollars every year; the National Cancer Institute budget for 2019 alone was $5.74 billion. If you add in charities and other funding, including pharmaceuticals, the amount spent on cancer research and programs likely exceeds $20 billion per year. There was no lack of cancer researchers. At the time of the publication of this book, PubMed.gov, part of the U.S. National Library of Medicine, listed 3.83 million articles published on the topic of cancer. *Three point eighty-three million* articles! But despite year after year of prodigious research, and a considerable investment of money and time, the common cancers in the early 2000s were just as deadly as they were in 1971. We were obviously losing Nixon's war on cancer. No, it wasn't lack of money or researchers that was the problem. It was the lack of fresh ideas.

After such an auspicious start, early twenty-first-century progress in cancer treatment ground to a virtual halt. The miracle drugs that had once seemed to promise a cure looked increasingly like exceptions—useful treatments for some that didn't impact most patients.

The Human Genome Project, completed in 2000, mapped an entire human genome but failed to shed significant light on the cancer problem, so in 2005, a new, sweeping genetic project was proposed. The Cancer Genome Atlas (TCGA) was even more ambitious in scope than the Human Genome Project. Instead of sequencing a single human genome, it would sequence hundreds of genomes from patients with cancer. The estimated cost for this project was $1.35 billion over nine years.[13] Many researchers believed that TCGA would be exactly the long-awaited cancer moon shot that would allow us to know the enemy and bring the fight to our own terms. Where the Human Genome Project sequenced a single human genome, TCGA would sequence more than ten thousand complete genomes.

Not everybody was convinced that TCGA was a good idea for those billions of research dollars. After all, we'd just finished one genetic megaproject, with little to show for it. Dr. Craig Venter, who had just recently completed the Human Genome Project, opined that "Diverting a billion or two dollars from other areas of research when it's not clear what answer we'd get, there might be better ways to move cancer research forward."[14] Prophetic, yes. Heeded, no.

Other researchers suggested that we should "Strap [ourselves] in and get ready for some serious 'more of the same.'"[15] The point, not well appreciated at the time, was that this megaproject was only the ultimate culmination and continuation of a futile line of research that had so far won us a few small battles in a war we were so clearly losing.

If the somatic mutation theory was not the answer to beating cancer, then sinking even more billions of dollars into dead-end research would sap resources from other, potentially more productive avenues. The cost of delaying this research would be measured in human lives. But SMT was the dominant theory, and dissenting opinions were not met with enthusiasm. In the end, even though the genetic paradigm had revealed few of cancer's secrets, the medical and scientific communities continued to invest billions more dollars into this failed strategy.

That's where we stood in 2006, poised to spend mega bucks on another mega cancer moon shot based solely on the waning legitimacy of SMT. In 2009, the Cancer Genome Atlas received another $100 million from the National Institutes of Health and an additional $175 million from the U.S. government in stimulus funding. TCGA eventually expanded to the larger International Cancer Genome Consortium, involving 16 nations, including more than 150 researchers over dozens of research institutes across North America.

By 2018, the PanCancer Atlas,[16] the culmination of TCGA and a follow-up research effort, was declared complete. A detailed genomic analysis of more than ten thousand tumors spanning thirty-three different cancer types had been successfully mapped. All the genetic codes for almost every cancer that afflicts mankind were now known.

Little excitement greeted this announcement, and few outside the cancer research community even knew about it. To be fair, few even *inside* the cancer research community cared. There were no headline articles in the *New York Times*. There was no cover story in *Time* magazine. It was almost as if we'd landed people on Mars and nobody took notice. What happened? The TCGA megaproject, like the HGP before it, failed to produce many useful insights into cancer, and worse, it produced no useful treatments.

The SMT had all seemed to fit so tightly together when we started. What went wrong? The problem was not that we had failed to find the genetic mutations. The problem was that we had found too many mutations—way, way too many.

THE PROCRUSTEAN BED

In ancient Greek mythology, Procrustes is the son of Poseidon, god of the sea. He would invite passersby to rest at his house for the night. When showing them their bed, if the guest was too tall, he would chop off their limbs until the bed fit just right. If they were too short, he would stretch them on a rack until the bed fit just right. The great contemporary thinker and philosopher Nassim Nicholas Taleb often uses this allegory of the Procrustean bed to describe how facts are often tortured to fit a certain narrative. The widely and often blindly followed somatic mutation theory of cancer required a Procrustean bed to fit the facts, too.

SMT considers cancer as a disease of randomly accumulated genetic mutations. But how many? As we've noted, some cancers are driven by a single mutation, but for most common cancers, this is far too simplistic. To fit these new facts into the existing Procrustean bed, cancer researchers now suggested the "two-hit hypothesis." Rather than a single mutation, there needed to be a combination of two mutations working together to produce a cancer. When this was recognized as also too simplistic, the SMT was expanded so that perhaps three or four mutations working in conjunction could produce the excessive growth and other features needed for a cell to become cancerous. Each accumulated mutation moves a normal cell closer and closer to becoming cancer. So, the real question now became: how many genetic mutations were needed for cancerous transformation? Two? Three? Four?

By 2006, there were unsettling signs that cancer mutations were more complex than first imagined. Much, much more. Cancer researcher Dr. Bert Vogelstein at Johns Hopkins found 189 genes that were significantly mutated in two of the most common solid cancers, breast and colon. There were not two or three or four gene mutations, there were *hundreds*. Worse, these genetic mutations are distinct from one cancer to the next. Each tumor carried an average of eleven mutations. For example, breast cancer number one would have eleven genetic mutations, but breast cancer number two would have eleven *completely different* mutations. Two breast cancers that looked identical clinically could be completely different genetically, barely resembling each other.[17]

These findings contrasted starkly with the hope offered by the Philadelphia chromosome, where every case of a specific cancer type shared a single identical mutation. If we chose one hundred random cases of chronic myelogenous leukemia (CML), almost all would bear evidence of the same Philadelphia chromosome. But for most common cancers, one hundred random cases had one hundred different genetic profiles of mutations, each completely distinct from the others.

It turned out that common cancers were *way* more complex than those previously studied. Not only were there more mutations, but those mutations varied from person to person. This meant that a drug that was effective for colorectal cancer in person A would not likely be effective for person B. Using the same "personalized, targeted medicine" principles as imatinib would require ten to twenty different "smart bomb" medications for any individual patient. The combinations are practically infinite.

As the first reams of data from the TCGA juggernaut started trickling in, the first inklings of the enormity of the challenge started to percolate. Instead of a few mutations, most cancers had fifty to eighty mutations. It was genetic pandemonium.

By the time 2015 rolled around, researchers had identified *ten million* different mutations. Ten million.[18] The mutations differed not only from patient to patient, but even within the same damned tumor in the same patient! They weren't trying to find a needle in a haystack. They were trying to find one particular needle in a needlestack, a much more painful research method.

More Procrustean ad hoc revisions to the SMT were needed. Multiple individual mutations were lumped into mutation "pathways." Those deemed important for carcinogenesis were called "driver" mutations. Other mutations, believed to have no effect, were called "passenger" mutations. These latter, all of a sudden, didn't count. Other researchers tried to divide these mutations between "mountains" and "hills." Mountains were mutations that were carried by the majority of cancers of that type. Hills were mutations carried by only a minority. Researchers were desperately trying to make sense of the overwhelming number of genetic mutations being identified.

Even with all this Procrustean work, studies estimated that each breast or colon cancer still required about 13 driver mutations.[19] In metastatic pancreatic cancer, 49 mutations were needed.[20] In 2013, Bert Vogelstein estimated that more than 140 different gene mutations could drive cancer growth and that each driver mutation increased the selective growth advantage of a cell by only a minuscule 0.4 percent.[21] A single gene mutation did not drive cancer. Most cancers had dozens and dozens of mutations that, individually, contributed only minimally to cancer growth. A more recent 2015 analysis of more than two thousand samples of breast cancer found more than 40 mutations in driver genes.[22]

Different cancers had different rates of mutation. Some cancers had hundreds of mutations, and some had none at all. A study of 210 human cancers found more than 1,000 different mutations. But a full 73 cancers had no identifiable mutations at all! This is obviously a problem for the SMT, which suggests that mutations drive cancer. If mutations caused cancer, how could 35 percent of the cancers in this study not have a single mutation? A full 120 different driver mutations were identified.[23] Lung cancer and melanoma of the skin contained almost 200 mutations per tumor, breast cancer was closer to 50, and acute leukemias closer to 10.[24]

The other significant problem for the SMT was the idea that all cancerous cells are cloned from an original cancer cell. All the cancer cells in a specific patient should have been genetic replicas of the original, but this clearly is not true, either. Within the same patient, metastatic cancer differs genetically from the original tumor. One site of metastasis may differ from another by twenty or more genetic alterations.[25] This degree of genetic heterogeneity was completely unexpected. Even within the same tumor mass, cells carried different mutations. Cancers differed with respect to their gene mutations in the following ways:

- Different types of cancer had different mutations.
- The same type of cancer in different patients had different mutations.
- The same cancer in the same patient had different mutations in the primary tumor compared to the metastatic cancer.
- The same cancer in the same patient had different mutations at different sites of metastasis.
- The same cancer in the same patients in the same tumor mass had different mutations.

The hard truth became one the research community couldn't deny: cancers were far, far more different genetically than they were alike. The quest to find the genetic mutations of cancer was *too* successful.[26] Cancer was a baffling mishmash of genetic peculiarities that had almost no connection to one another. Genetic mutations were everywhere and nowhere. Some cancers had hundreds of mutations, and others had none at all. The rate of mutation necessary to develop a cancer is much higher than the known rate of mutation in human cells. Normal cells just don't mutate anywhere close to the rate needed to produce cancer. Cracks had formed in the genetic paradigm, and mutations were invading it like weeds.

8

THE DENOMINATOR PROBLEM

BY THE 2000s, hundreds of potential cancer-causing genes had been identified. Everywhere researchers looked, there were more oncogenes and tumor suppressor genes. Presumably, a single mutation in any one of the normal growth-controlling genes could cause cancer. So why wasn't everybody getting cancer?

There is a common issue in surveillance studies called the denominator problem. Suppose we analyze one hundred great baseball players and discover that every one of them has a liver. We might conclude that having a liver makes you a great baseball player. But this is a logical fallacy, because many people with livers are *not* great baseball players. This is the denominator problem. Out of all people with a liver, how many are great baseball players compared to the many who are not?

If we take one hundred cancer samples and find that all one hundred have genetic mutations, we might conclude that having genetic mutations is the key to developing cancer. But that conclusion is not logically warranted because we are still missing an important piece of information: the denominator. If one hundred samples of *noncancerous* tissue also contain genetic mutations, this clearly diminishes the importance of genetic mutations in the genesis of cancer. To assess the importance of genetic mutations for cancer, we need to compare the following:

Without knowing the denominator, we could not know the importance of genetic mutations. How many cells had genetic mutations but were not cancer? This number turned out to be pretty high. A full 4 percent of a cell's DNA could have mutations, and that cell could still look and act normally. This is a remarkably high degree of tolerance.[1] If we estimate that there are twenty-five thousand genes in a human, then you could have roughly one thousand gene mutations without developing cancer.

$$\frac{\text{Genetic mutations in cancerous tissue}}{\text{Genetic mutations in noncancerous tissue}}$$

Cancer patients had lots of mutations, yes, but so did cancer-free people. Some healthy people even had the *identical genetic mutations* as those with cancer. A 2012 detailed analysis of 31,717 cancer cases concluded that "the vast majority of, if not all, aberrations that were observed in the cancer-affected cohort were also seen in the cancer-free subjects."[2] These discoveries posed a real problem for the SMT model. The Cancer Genome Atlas pinpointed plenty of mutations, but the important question *not* asked was: how many normal cells carry the very same mutations in the cancerous pathways but do *not* develop into cancer? The SMT predicted that normal cells would carry few or no critical mutations. That prediction was way off. Recent genetic sequencing of discarded skin from cosmetic surgeries, completely free of cancer, has yielded astounding results. Almost one in four samples contained mutations known to be involved in cancer—and yet no cancer was present.[3]

In addition, genetic sequencing of esophageal specimens taken from organ donors with no prior history of cancer has also yielded more stunning results. The average *healthy* cell with no evidence of illness or cancer contained "at least several hundred mutations per cell in people in their twenties, rising to over 2,000 mutations per cell late in life."[4] The *NOTCH1* oncogene, for example, considered to be a prime driver mutation of esophageal cancer—it is present in about 10 percent of all such cancers—was also found to be present in up to 80 percent of all esophageal cells in patients free of cancer. Eighty percent! In middle-aged and older specimens, 2,055 mutations were found in that single oncogene alone—yet none developed into cancer. In other words, the seed was there, but cancer had not grown.

This evidence points to a simple yet stunning fact: the premise that a single mutation in an oncogene or tumor suppressor gene is the origin of most cancer growth is far too simplistic. The SMT had ignored the denominator problem. But there was another problem still. The genetic mutations were a proximate, rather than a root, cause of cancer.

PROXIMATE VERSUS ROOT CAUSES

For any disease, understanding the root cause (known as the etiology) is the basis for all rational treatment. Between root cause and final outcome are many intermediate steps called proximate causes. These are usually immediately obvious. The root cause of disease is what we generally consider the "real" cause of the event in the first place, and it often requires higher-level thinking to determine.

For example, liver failure is caused by fibrotic scarring known as cirrhosis. This information is true, but not very useful. Rather, we want to know what causes the *cirrhosis*. If you know that the cirrhosis is caused by the hepatitis C virus, then you can prescribe an antiviral medication to treat it. If you know that the cirrhosis is caused by alcohol, then you can advise a patient against drinking alcohol. Liver failure is caused by cirrhosis, true enough, but cirrhosis is only the proximate cause. Successful treatment depends upon knowing the ultimate, or root, cause (see Figure 8.1).

This approach applies to most problem-solving, not just in medicine. For example, an airplane crashes when the force of gravity exceeds the lifting force. You might conclude from this simplistic analysis that the key to preventing plane crashes is to

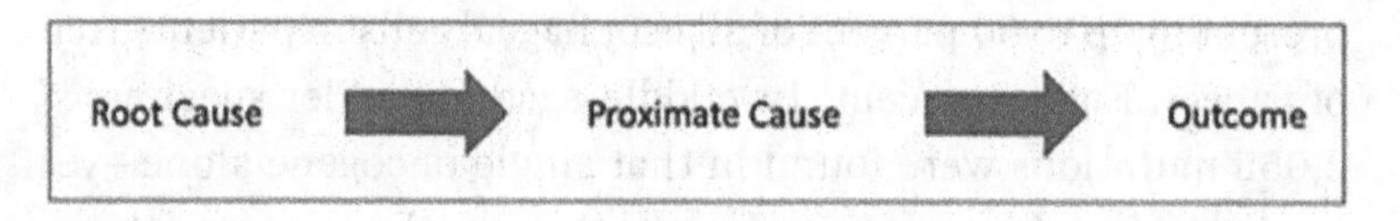

Figure 8.1

increase lift by building bigger wings, or decrease gravity by decreasing weight. Both solutions are virtually useless because they address proximate, not ultimate, causes.

To understand the problem, you must go one level higher in logical thinking. *Why* did the force of gravity exceed the lifting force? We could catalogue all the parts of the airplane that malfunctioned and why: cracks in the wing, cracks in the tail, engine failure, electrical failure. This is *how* the plane crashed (proximate cause), but it still does not solve the question of *why*. Why did all these problems occur? The root cause may actually be poor maintenance, poor pilot training, inclement weather, or any number of other root causes.

A solution targeted at the root cause, such as better airplane maintenance, better pilot training, or better weather forecasting, is highly effective. Bigger wings, lower weight, or larger engines are not. The higher-level analysis of the cause provides effective solutions. Whereas treating the root cause is successful, treating the proximate cause is not (see Figure 8.2). How does this apply to cancer? Genetic mutations were only the proximate cause of cancer. What was driving those mutations to occur?

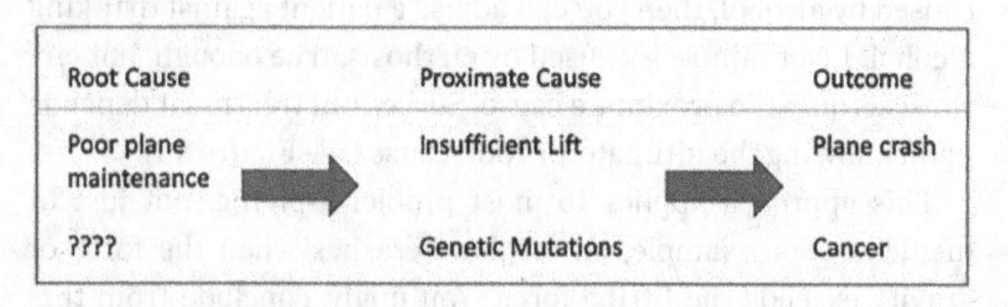

Root Cause
Proximate Cause
Outcome
Poor plane maintenance
Insufficient Lift
Plane crash
????
Genetic Mutations
Cancer

Figure 8.2

In cancer research, we spent enormous resources to catalogue thousands of different mutations. Lung cancer, for example, has mutations in the oncogenes *AKT1, ALK, BFAF, EGFR, HER2, KRAS*, and *NRAS*. This is *how* the cancer developed, but not *why*. These are all proximate causes, but not the root cause. If I ask you, "What causes lung cancer?" Would you answer with "Lung cancer is caused by a mutation in the *AKT1* gene" or "Lung cancer is caused by smoking"? It doesn't even much matter what genetic mutations are induced by cigarette smoking if we know the root cause. I can save many more lives by knowing that smoking causes cancer than by knowing all the various mutations seen in lung cancer. Smoking cessation is one of the most successful anticancer measures we have at our disposal.

Once we identified occupational exposures as a root cause of cancer (e.g., soot, asbestos), prevention dramatically reduced risk. Once we identified viral and infectious causes of cancer, prevention (e.g., vaccinations against hepatitis B or HPV, improved sanitation) reduced risk. But we must find the root cause, not the specific genetic mutations of liver, cervical, or stomach cancer. In almost all human disease, treating the root cause, not the proximate cause, is the key to success.

Cancers contain many mutations; there is no doubt about that. We've spent decades cataloguing all of them. But *why* does cancer develop these mutations? What is the driving force behind these mutations? Cancer paradigm 2.0 suggested that these mutations were accumulated purely by chance (see Figure 8.3). The known root causes of cancer (chemicals, radiation, and viruses) increase mutation rates, allowing some to randomly aggregate into cancer. Again, an infinite number of monkeys randomly pounding the keys of an infinite number of typewriters will eventually produce *War and Peace*.

This is one of the fatal flaws of the somatic mutation theory. Mutations were accumulating, *but their occurrence was anything but random*. Given that hundreds of mutations were working

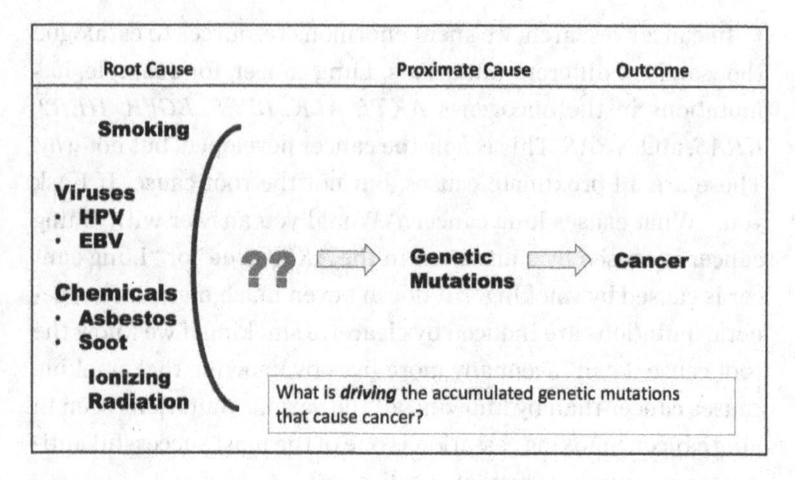

Figure 8.3

closely together to create cancer, those mutations looked purposeful and coordinated.

Cells are like finely tuned watches. Everything has a purpose. Randomly removing a screw is unlikely to make a watch work better. Rather, the more probable outcome is malfunction. In a cell, a random mutation is most likely going to be harmful or even lethal. Thus, the odds of assembling, just by chance, two hundred random mutations that didn't kill the cell but instead bestowed powerful new abilities on it are somewhat less than the chances of my winning the Powerball lottery.

The rate of gene mutation in the general population is very low and much, much lower than the rate necessary to drive so many cancers. The process of randomly assembling fifty to two hundred genetic mutations together in a coherent form is so extraordinarily complex that cancer should be a vanishingly rare disease. But cancer is not a rare disease. Instead, it is exceedingly common, the number two killer of Americans. An estimated 50 percent of the general population will have colonic adenomas (precancerous lesions) by the age of eighty.[5] Some 80 percent of men over age ninety will exhibit evidence of prostate cancer.[6] Breast cancer is estimated to have a lifetime risk of one in nine women.

There is yet another fatal flaw in the SMT. If all the mutations of cancer are accumulated randomly, then why do all cancers share so many of the *same* hallmarks? To become cancer, cells must gain a number of special new abilities. We've discussed the four hallmarks of cancer: it grows, becomes immortal, moves around, and uses the Warburg effect. How can every cancer in history randomly develop all these wondrous abilities from scratch? How do two hundred different random mutations somehow still give the same end result? If those typing monkeys produce one hundred copies of *War and Peace* but no other classic novel, then it is not random. Those monkeys were *trying* to write *War and Peace*. The cell is *trying* to develop into cancer.

If one airline suffers hundreds of crashes every year, where other airlines have none, then there is nothing random about it. In cancer, asbestos causes DNA damage, genetic mutations, and also a cancer called mesothelioma, but not breast or colorectal cancer. Almost nothing else in the world causes mesothelioma. So, the genetic damage caused by asbestos is clearly not random.

These hallmarks of cancer are carefully selected. Something is pushing these oncogenes and tumor suppressor gene mutations toward growth, movement, immortality, and the Warburg effect. The next great leap forward in cancer paradigms would be to understand what drives these changes.

PREPOSTEROUS REDUCTIONISM

The somatic mutation theory is simple, compelling, elegant, and largely incorrect. In 2002, cancer researchers Hahn and Weinberg published an article in the *New England Journal of Medicine* observing, "The actual course of research on the molecular basis of cancer has been largely disappointing. Rather than revealing a small number of genetic and biochemical determinants operating within cancer cells, molecular analyses of human cancers have revealed a bewilderingly complex array of such factors."[7] The next paradigm of cancer would need to make sense of this bewildering array.

The ultimate test of a cancer paradigm's usefulness is its ability to develop revolutionary treatments. The SMT had started with great promise, delivering the remarkable drugs imatinib and trastuzumab, but these breakthroughs proved to be the exceptions, not the rule. Since then, successful treatments derived from the genetic paradigm of cancer have virtually ground to a halt; a generous estimate of the number of useful genetically targeted medications is five. I think we can all agree that five drugs in fifty years of genetics research is hardly winning the war on cancer.

There was yet another problem with developing genetically targeted medications for cancer: drug resistance. If the genetic target constantly changes, how can you possibly develop an effective therapy? As soon as you targeted one pathway, the cancer would simply find another way to get around it. Cancer could activate different genes to bypass whatever pathway we tried to block. We were attacking cancer's strength, not its weakness. Cancer often contains hundreds of mutations. Blocking a single mutation is unlikely to successfully halt growth because there are ninety-nine other mutations. Cancer can mutate around whatever we throw in its path, because that's what cancer does better than anything else.

Genetic mutations may explain the mechanism of *how* cancers keep growing, but they do not explain the fundamental question of *why* these genes mutated. The SMT fails because it is entirely inward-looking, toward our genes, instead of outward-looking, toward the environment. But so many different environmental attributes obviously affect cancer risk. The seed is important, but the soil matters more.

Cancer paradigm 2.0 treats cancer as a genetic lottery, but cancer is not merely a matter of bad luck. Most of the known risk of cancer is due to environmental factors, not genes. It's tobacco smoke. It's radiation. It's infections. It's the last great frontier of cancer medicine, one that is little explored: diet. The good news is that these factors are largely within our control.

The SMT was a textbook case of not seeing the forest for the trees. When you're standing in the middle of a forest, it doesn't seem so great. Here's a tree. Here's another tree. Here's a third tree. What's the big deal? But if you see a national forest like Yosemite from a helicopter, you can appreciate its stunning beauty.

Consider another analogy. You are studying the significance of the Declaration of Independence, so you closely examine each individual letter, just as cancer researchers might look very closely at each new genetic mutation. You create a "Complete Letter Atlas," just as cancer researchers created a Cancer Genome Atlas. The letters *A*, *E*, and *T* appear (hypothetically) several hundreds of times, but *Z* and *X* barely at all. Does this exhaustive effort help you understand the role of an independent America in world history? Not at all.

In science, this has been called "preposterous reductionism." Reducing a problem to its smallest component necessarily means missing the big picture. You cannot understand rush-hour traffic by painstakingly cataloguing how each driver braked and accelerated. Yes, the combination of those individual acts of braking and accelerating caused the traffic jam. But knowing this level of detail is not useful. These were not random acts. *Why* were those drivers braking and accelerating?

Similarly, you cannot understand cancer by painstakingly cataloguing thousands of genetic mutations in tumor suppressor genes (brakes) and oncogenes (accelerators). Yes, those individual mutations together caused cancer. But knowing this level of detail is not generally useful.

With the SMT, we zoomed into the cancer problem too closely—right down to its genetic makeup—and found gibberish. We found thousands of different mutations spread among hundreds of oncogenes and tumor suppressor genes, but that didn't help piece together the story as a whole. Each mutation describes a tiny, tiny piece of the puzzle, and the relentless focus on cataloguing new mutations drained the lifeblood from other areas of cancer research.

Insisting that cancer is a disease of collected genetic mutations is like insisting that the Declaration of Independence is a collection of letters. True enough, but so what? How does it help us understand cancer?

CONCLUSION

The somatic mutation theory did advance our understanding of cancer, but not in the ways we'd anticipated. Instead of decoding the genomic component, it unearthed the disorienting number of genetic mutations involved. A single colorectal cancer contains about one hundred separate genetic mutations, and those mutational patterns differ greatly from one patient to the next.[8] Other studies estimated that there were eleven thousand mutations in human colorectal cancers.[9] The SMT tried to adjust for these unexpected findings by adding a series of ad hoc modifications. Some mutations, termed "drivers," mattered, and others, termed "passengers," did not. There was not a single genetic clone, but changes in clonal evolution over time. On and on it went. With each iteration, the complexity of the SMT increased, until it was no longer a simple, elegant theory. Instead, it sported multiple unwieldy, cobbled-together additions.

Finally, the entire SMT collapsed under the weight of all these modifications. It simply couldn't be altered enough to fit the known facts of cancer. Worse, it failed to deliver more than a few effective treatments. In the meantime, cancer therapy lagged far behind almost every other area of medicine, and patients died. It was time to leave the Procrustean bed behind.

9

A FALSE DAWN

IMATINIB, THE FIRST drug of the age of personalized, precision cancer medicine, was a true game changer. With its introduction, a patient suffering from chronic myelogenous leukemia (CML) could reasonably expect to live as long and be as healthy as somebody without the disease.[1]

Before imatinib, a sixty-five-year-old man diagnosed with CML was expected to live less than five years, compared with fifteen years for somebody without the disease. With imatinib, that man with CML enjoyed a life expectancy virtually identical to that if he had not developed CML.

But other genetically targeted drugs, while effective, don't necessarily qualify as game changers. Such is the case with the anaplastic lymphoma kinase (*ALK*) inhibitor called crizotinib, which has been hailed as one of the biggest breakthroughs in genomic medicine of the last two decades. This drug has been proven to treat certain types of lung cancer (non-small-cell), but its benefits are limited. A recent meta-analysis of crizotinib found only scant evidence that the drug improved overall survival.[2] A game changer? Debatable. In 2019, a single month's worth of medication cost $19,589.30.[3]

Much of the low-hanging fruit of the genetics revolution had been plucked. More recently, developed drugs have offered diminishing returns. But despite this poor showing, researchers were slow to change course. Even as late as 2017, Dr. José Baselga, former physician in chief of Memorial Sloan Kettering Cancer Center, one of the premier cancer hospitals of the United States, pleaded for more money to continue the drive for what he called "genome driven oncology."[4] "Cancer is a disease of the genome," he stated baldly, desperately citing the now-ancient 1990s-era discovery of imatinib. Baselga himself was forced to step down in disgrace in 2018, when a *New York Times* investigation revealed that he had neglected to disclose financial conflicts of interest in a stunning 87 percent of articles he had written the previous year.[5]

The sexy idea of personalized, precision cancer treatment appeals widely to patients, physicians, and funding agencies alike.[6] In 2015, President Barack Obama, unable to resist the siren song, dedicated millions more dollars to the Precision Medicine Initiative Cohort Program. Even by then, though, the evidence was overwhelming that genetics-based precision medicine could not fulfill its initial lofty promise.[7]

Personalized, precision cancer medicine depends upon two crucial steps: detecting the patient-specific genetic mutation, and delivering a targeted drug for that mutation. We've succeeded at step one, having identified thousands of gene variants, far more than we can possibly investigate. But what about step two? Could we actually deliver a drug to target that mutation? In 2015, only eighty-three of two thousand consecutive patients with complete genome testing treated at the specialty cancer hospital the University of Texas MD Anderson Cancer Center, in Houston, were eventually matched to a targeted treatment, representing a paltry 4 percent success rate.[8]

The National Cancer Institute fared no better in its Molecular Analysis for Therapy Choice (NCI-MATCH) trial.[9] After mapping 795 cancer genomes, the NCI matched 2 percent to a targeted therapy—and not all those matched actually respond to treatment. Even at a highly optimistic response rate of 50 percent, this would constitute a 1 percent response rate to personalized cancer care, with an expected survival improvement measured in months.

This was the state-of-the-art treatment in cancer genomic medicine in 2018, and it really sucked.

The poor results were not for lack of available medications. The Food and Drug Administration had been busy approving a cornucopia of new "genome-driven" cancer medications at an unprecedented pace. From 2006 to 2018, thirty-one new drugs for advanced or metastatic cancer were approved. Sounds pretty amazing. Almost three new drugs a year were being mainlined into the sickest cancer patients. Yet, despite so many new medications and the rapidly advancing technology of cancer genome sequencing, a 2018 study estimates that only a minuscule 4.9 percent of patients actually derived some benefit from genome-targeted treatment.[10] Even after fifty years of intensive study, this paradigm of cancer fails more than 95 percent of people. Not good. How could so many "new" genomic medications deliver so few benefits?

One reason is that most "new" medicines are not new at all, but merely copycats of existing ones. Developing an innovative drug is hard work and carries substantial financial risk. Even highly effective drugs may fail due to unacceptable side effects. Copying existing drugs, rather than innovating new drugs, is a far more profitable strategy. If drug company A successfully develops a cancer drug that blocks gene target A, then at least five other drug companies will soon develop five other, almost identical drugs. To circumvent patent protection, they change a few molecules on a distant chemical side chain and call it a new drug. These copycat drugs carry almost no financial risk because they are virtually guaranteed to work.

Imagine that you sell children's books. You could either write an original novel, or simply plagiarize the entire Harry Potter series but change the name of your hero to "Henry Potter." Great novel? Yes. Makes money? Yes. Innovative? Not even a little. Hence the proliferation of drugs such as imatinib, nilotinib, and dasatinib, which are all basic variations on the same molecule.

Instead of finding new genetic treatment, all those research dollars of the big pharmaceuticals just gave us some serious "more of the same." Plagiarizing is a better corporate strategy than innovating. The benefits may be marginal, but the profits are maximal.

There are other ways to give the appearance of making progress. Of the many ways to game the system in medical research, using surrogate outcomes is one of the best.

SURROGATE OUTCOMES

Surrogate outcomes are results that, by themselves, are meaningless, but they predict outcomes we actually do care about. The danger of trusting surrogate outcomes is that they don't always reflect the desired outcome accurately. For example, two heart drugs (encainide and flecainide) were widely prescribed because they reduced extra heartbeats (called ventricular ectopy) after a heart attack, a surrogate outcome for sudden cardiac death. But a proper clinical trial proved that these two drugs significantly *increased* the risk of sudden death.[11] The drugs weren't saving patient lives; they were ending them.

In cancer trials, progression-free survival (PFS) and response rate (RR) are two commonly used surrogates for the outcome we are most interested in, overall survival. PFS is measured as the time from treatment to disease progression, defined as a less than 20 percent increase in tumor size. Response rate (RR), in this scenario, would be the percentage of patients whose tumor shrinks more than 30 percent. In order to be useful, these surrogates must predict the clinical outcome—overall survival—but they don't.[12] The overwhelming majority of studies (82 percent) find the correlation between surrogate markers and overall survival to be low.[13] Both PFS and RR are surrogate outcomes based entirely on reducing the size of a tumor, but surviving cancer depends almost entirely upon stopping metastasis, a fundamentally different beast

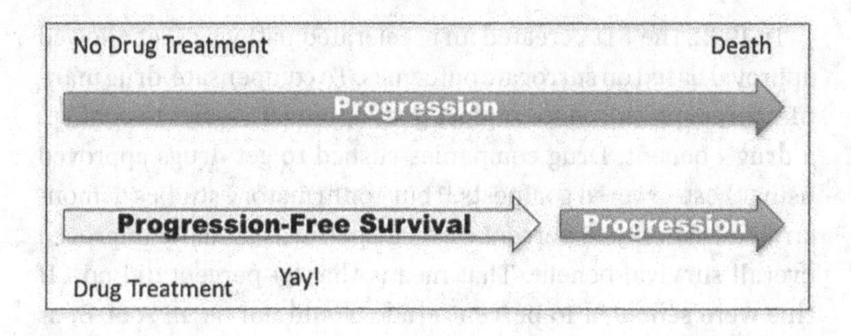

Figure 9.1

(see Figure 9.1). Just because an outcome is easily measurable does not make it meaningful.

Size is only a single factor contributing to a tumor's overall deadliness, and arguably, it is one of the least important. Cancer becomes more lethal when it mutates to become more aggressive or more likely to metastasize, and size-based surrogate outcomes like PFS and RR make little difference. Shrinking a tumor by 30 percent has virtually no effect on survival because of cancer's uncanny ability to regrow.

Surgery is never done to remove 30 percent of the tumor mass, because that would be futile. Surgeons go to extraordinary lengths to make sure they "got it all," because missing even a microscopic part of a tumor means that the cancer will definitely recur. While a minuscule 6 percent of drugs have reached this threshold of complete remission, from 2006 to 2018, fifty-nine oncology drugs were approved by the FDA on the basis of RR.

All these significant problems of using the cheaper but flawed surrogates for overall survival are well known. Before 1992, less than 3 percent of trials used surrogate outcomes. Then things changed. From 2009 to 2014, two thirds of FDA approvals came from trials using PFS as a surrogate outcome.[14] For cancer drugs given the "breakthrough" designation, 96 percent relied on a surrogate end point.[15] What happened?

In 1992, the FDA created an accelerated pathway that allowed approval based on surrogate outcomes. To compensate, drug manufacturers promised to conduct post-approval studies to confirm a drug's benefit. Drug companies rushed to get drugs approved using these lowered goalposts,[16] but confirmatory studies demonstrated that only 16 percent of drug approvals actually improved overall survival benefit. That means that 84 percent did not. If this were school, a 16 percent grade would not be an A or B, or even an F. It would be like submitting a term paper and receiving the grade H. It's bad. Very bad.[17]

This reliance on surrogate outcomes has led to costly mistakes already. In 2008, the FDA approved the drug bevacizumab for metastatic breast cancer based on a 5.9-month improvement in PFS,[18] even though overall survival was unchanged. Subsequent studies discovered *reduced* PFS, with no benefit to overall survival or quality of life, and substantial toxicity. The FDA withdrew approval for bevacizumab's use in treating breast cancer in 2011.[19]

The cautionary tale provided by bevacizumab would be unheeded time and again. In 2012, the drug everolimus was approved for the treatment of metastatic breast cancer based primarily on surrogate outcome studies.[20] By 2014, follow-up studies made clear that this drug provided no substantial benefits.[21] In 2015, the drug palbociclib received approval for use in breast cancer, but once again, later studies failed to find any survival benefits.[22] In the meantime, thousands of men and women suffered the devastation of holding out hope for a miracle cure, only to see it vanish before their eyes—their dreams of recovery slowly drained as surely as their bank accounts.

Surrogate outcomes allow earlier approval, which saves time. You would think this is a good thing for cancer patients, for whom time is precious. But how much time is actually saved? A modern cancer drug takes a median of 7.3 years to come to market,[23] with 38 percent approved on the basis of RR and 34 percent on the basis of PFS. The use of surrogate outcomes generally saves an estimated total of eleven months of time. Is this time savings really worth an error rate of over 80 percent?

Five cancer drugs have already been approved through the accelerated program and then withdrawn when subsequent studies ultimately showed them to be useless. These drugs were marketed to a vulnerable public for 3.4 to 11.5 years.[24] That's shameful. Imagine that you sold your house to afford your cancer treatment, not knowing that the drug being hailed as the latest and greatest was literally useless. Worse, undergoing treatment with that drug meant that you weren't able to receive treatments that might otherwise have had some benefit.

Despite the seventy-two "new" cancer medications approved between 2002 and 2014, the average drug extended life by only an average of 2.1 months.[25] That's only an average, and most drugs offered no survival benefits at all. The sobering reality of the poor performance of new cancer drugs[26] is at striking odds with the public perception that the medical community is making leaps forward in the war against cancer.[27] One study found that half the drugs hailed as "game changers" in the media hadn't yet received FDA approval for use; 14 percent of these overhyped drugs had not even been tested in humans. Scientists have cured many thousands of rat cancers. Human cancers, though, not so much. Cancer research produces lots of publicity and excitement, but little actual advancement. Breakthroughs are heartbreakingly rare.

The ultimate goal of cancer treatment is to improve overall survival and quality of life. These patient-centered outcomes are hard to achieve and expensive to measure. To show benefits where none truly existed, you can politely move the goalposts by using surrogate end points.[28] For a pharmaceutical company, a positive study means FDA approval, which means revenue. But many of the drugs in use are of limited effectiveness, so what is a pharmaceutical company left to do? Why, raise prices, of course!

RAISING PRICES

When it was launched in 2001, imatinib cost $26,400 per year. A steep price, to be sure, but it was truly a miracle drug and worth every penny. By 2003, its sales totaled $4.7 billion worldwide—a mega blockbuster generating huge, well-deserved returns for the drug developer. Prices (inflation adjusted) began to edge higher in 2005, rising by about 5 percent per year. By 2010, prices were soaring by 10 percent per year above inflation.[29] Adding further to the bottom line, many, many more patients were living longer with their disease. This was a double bonanza for Big Pharma.

- More patients surviving CML = more customers.
- More customers + higher prices per patient = more money.

By 2016, a year's worth of this miracle drug cost more than $120,000. By this time, the drug had already been on the market for fifteen years. That's ancient history in medical science. It wasn't cutting-edge stuff anymore; it was medical school stuff. The actual cost of manufacture, even after adding a 50 percent profit margin, is estimated at $216 a year.

When new competitors to imatinib emerged, prices should have dropped. But a strange thing happened: prices increased. Price competition is not nearly as profitable as price fixing and collusion, so prices continued their ascent into the stratosphere. Dasatinib, a copycat of imatinib, was priced higher than the drug it was trying to replace: it was an iPhone knockoff priced higher than a genuine iPhone.[30] This exerted a strong pull on imatinib's price: upward. Drug costs were limited only by what the payer (mostly the taxpayer) could bear.

Raising drug prices after launch is now commonplace. On average, inflation-adjusted prices rise 18 percent in the eight years after launch[31] regardless of competition or efficacy. Imagine if Apple sold the original iPhone each year with no upgrades but raised the price annually by 18 percent—who would buy a new phone? No one. But cancer patients don't have the luxury of making such a choice. As a result, price gouging is a routine part of today's cancer landscape.

In the late 1990s, paclitaxel became the first blockbuster anti-cancer drug to reach $1 billion in sales.[32] By 2017, a drug would need to rack up sales of $2.51 billion just to crack the top ten of oncology drugs.[33] That is the reason that cancer medications took three of the top five spots on the list of the best-selling drugs of 2017.[34]

The top-selling drug of 2017 was Revlimid, a derivative of thalidomide, with sales of $8.19 billion. Such sales are easy to achieve when a drug costs in excess of $28,000 per month. Introduced in the late 1950s, thalidomide was notoriously prescribed for the treatment of morning sickness during pregnancy. Tragically, its use in pregnant women caused death and limb deformation in fetuses, and it was forced off the market in 1961. Thalidomide was reborn, though, in 1998, when it was approved for the treatment of leprosy and, more excitingly, multiple myeloma, a type of blood cancer.[35]

In the 1950s, the drug cost pennies. In 1998, the reborn thalidomide cost $6 per capsule. Only six years later, that price had risen almost fivefold, to $29 per capsule. The cost of making the drug is minuscule. In Brazil, a government lab sells it for $0.07 each.[36] The average annual cost for a cancer drug prior to 2000 was less than $10,000. By 2005, this figure had reached between $30,000 and $50,000. In 2012, twelve of thirteen new cancer drugs approved were priced above $100,000 per year. That tenfold increase in twelve years is far in excess of reasonable.[37]

Combining the high prices of cancer medications with low efficacy means that the cost-effectiveness ratio is off-the-charts bad. A generally acceptable cost of one quality-adjusted life year (QALY) is $50,000.[38] Cervical cancer screening has an estimated cost per QALY of less than $35,000.[39] Imatinib was starting to push the limits, at $71,000. But regorafenib, used in the treatment of metastatic colorectal cancer,[40] costs a staggering $900,000 per QALY.

Price is a reasonable heuristic for quality for most consumer goods. Expensive stuff is generally of higher quality. Nike shoes are generally more expensive and of higher quality than dollar-store shoes. This does not apply to cancer medicine, where a high-priced drug does not necessarily work better than a cheaper drug. Many expensive drugs may not even work at all.[41] This is obviously a big problem when drug costs are the single largest cause of personal bankruptcy in the United States.[42]

LOSING THE WAR

Cancer paradigm 2.0 had hit rock bottom. Cancer was still undefeated, and the situation looked bleak. Millions of cancer research dollars over several decades produced an abundance of new drugs. Some are truly great, but most are marginally effective yet fantastically expensive. Benefits are borderline, toxicity is high, and cost is higher still. The drugs weren't particularly useful, but they *were* particularly profitable. Lack of new drugs, the use of surrogate outcomes, and sky-high-but-still-rising prices: that's how you lose the war on cancer. But the day is always darkest just before the dawn.

PART III

TRANSFORMATION

(Cancer Paradigm 3.0)

10

THE SEED AND THE SOIL

IT WAS THE English surgeon Stephen Paget (1855–1926) who first compared cancer to a seed. In 1889, he wrote, "[S]eeds are carried in all directions; but they can only live and grow if they fall on congenial soil."[1] Plants grow when the seed, the soil, and the conditions are amendable to growth. If any one component is off, the plant does not grow. Cancer cells, too, are seeds full of malignant potential; but without the appropriate soil, they rarely bloom.

Without a seed, a plant will not grow no matter the soil or the conditions. A viable seed planted in clay will not grow. A viable seed planted in the right soil will still not grow without adequate light and water. You must have the right seed, the right soil, and the right conditions (environment). Cancer, too, is a seed that flourishes in the right soil, under the right conditions. Unfortunately, cancer research up until this point has focused exhaustively and almost exclusively on the seed (gene mutations) while largely ignoring the soil and conditions.

Consider another example. Some of the best ice hockey players in the world come from Canada, and some of the best basketball players come from the United States. If we were looking only at the seed in each of these environments, we might hypothesize that Canadians and Americans possess genetically distinct attributes: one with a "hockey" gene and the other with a "basketball" gene. That's obviously incorrect. The difference in skill and accomplishment is largely the result of different environments and cultures. Seeing a "soil" problem purely as a "seed" problem is a critical mistake.

The genetics of cervical cancer are far less important than the presence of the human papillomavirus (HPV). The genetics of lung cancer are far less important than exposure to cigarette smoke. The genetics of breast cancer are far less important than the lifestyle differences between Japan and America. The genetics of mesothelioma are far less important than the presence of asbestos in an environment. The genetics of stomach cancer are far less important than a positive test for *H. pylori.* The list, of course, goes on and on. So much of what we know about the etiology of cancer is derived from looking at soil problems, not seed problems.

And yet, the somatic mutation theory (SMT) focuses inward, on the "seed" problem. It is true that in certain rare cases, the seed is the most important factor in a cancer. The abnormal Philadelphia chromosome is the major cause of chronic myelogenous leukemia (CML). Fixing the genetic seed problem with imatinib largely cures the disease. It doesn't much matter if patients smoke, get viral infections, or move from Japan to America. If you have the Philadelphia chromosome, you will likely get CML. Unfortunately, these types of cancer are exceptions, not the general rule. For most cancers, studying the seed alone does not help you understand why it grows.

If a Japanese woman moves to America, her risk of developing breast cancer roughly triples within two generations. The genetic seed is the same, but the soil is different. While the information is alarming, it is also empowering: it means that if we can understand what kind of soil breast cancer needs, we can potentially reduce the risk by one third simply by changing that environment, predominantly through modifications to diet and lifestyle. This is an incredible opportunity because it means that our genes are not our destiny.

EPIGENETICS

The emerging field dedicated to the study of how an environment can change an organism without alterations to its DNA is called epigenetics. The word *epigenetics* is derived from the Greek prefix *epi-*, which means "over" or "above." Gene regulation occurs at the level above the DNA, hence the name *epigenetics*. Epigenetics is not concerned with the genetic changes or mutations encoded into the DNA, but how those genes are expressed or not.

Epigenetics affects the packaging of the genes rather than the genes themselves. The details of this process are outside the scope of this book, but the simple version is this: one of the main mechanisms of epigenetic changes is known as DNA methylation. Changes to the DNA methylation of tumor suppressor genes can silence those genes,[2] which favors growth and cancer. This change to gene expression, and therefore risk of cancer, occurs without any genetic mutation.

Think of the sheet music for a song. The notes provide a blueprint, but you may attach crescendos, decrescendos, and other markers of how to play the song to alter it in a hundred ways. The very same written notes can produce very different songs. A Beatles song sounds completely different if sung by Aerosmith. In the case of genes, the DNA sequence of a gene provides the blueprints, but the environment can alter the expression of the genes in a hundred ways, without any change to the underlying blueprint. A gene mutation, however, is a permanent change, akin to inserting or deleting notes from the sheet music.

Many environmental factors, such as diet and exercise, can influence gene expression. Epigenetics upends the old dogma that the genetic code is the key determinant of cellular gene expression and function. The packaging of the gene may be as important as, or even *more* important than, the gene itself, and these epigenetic changes are predominantly influenced by environmental factors.

This phenomenon obviously challenges the old SMT model, which focuses purely on gene mutations.

If the genes themselves are unchanged, then deciphering the underlying genetic code at great expense is of limited usefulness. By the time work on the Cancer Genome Atlas (TCGA) started, it was already well known that changes in DNA methylation are vital to the development of some cancers.[3] A number of known carcinogens are considered to act through epigenetic pathways. In colon cancer, up to 10 percent of protein-coding genes are methylated differently from normal colon cells, emphasizing the role of epigenetics.[4]

This represents a huge shift from the SMT. Developing cancer depends upon both the intrinsic mutations and the extrinsic selection pressures of the environment in which that cancer grows. This does not minimize the importance of having the right seed, but instead enhances our understanding of cancer's growth to include the soil. The environment exerts natural selection pressure on those "seeds" best suited to survive. A cancer may bloom or stay dormant depending upon the state of the body.

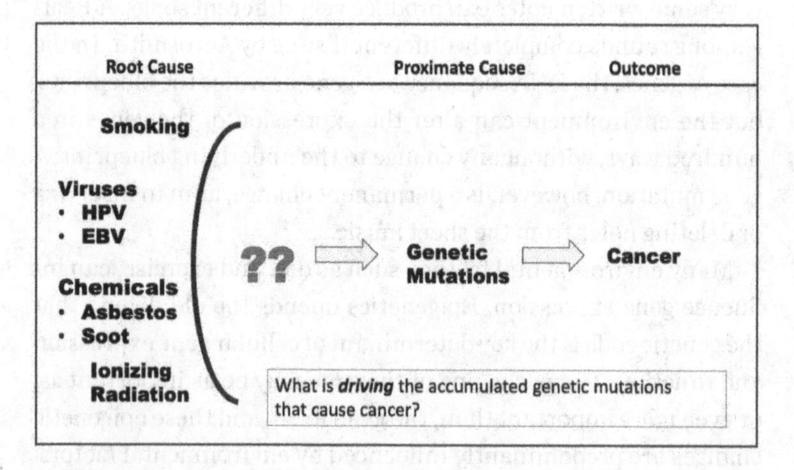

Figure 10.1

This new paradigm offers a more nuanced understanding of how cancer cells interact with their environment to produce clinically meaningful cancers. The environment selects certain seeds to thrive and others to wilt.[5] What is driving that selection? That is the real question (see Figure 10.1).

DEVELOPING NEW PARADIGMS

In 2009, the National Cancer Institute (NCI), in an uncharacteristic move, reached out beyond the expected cadre of researchers to ask other scientists for help in the war on cancer. The call went out not to cancer biologists or cancer researchers, but to theoretical physicist Paul Davies and astrobiologist Charley Lineweaver. With no prior knowledge of cancer and, most important, no preconceived notions, these two would usher in the next chapter in our understanding of cancer.[6]

The NCI perceptively grasped that funding the same researchers gets you the same tired, and ultimately not very useful, answers. But physicists could offer an entirely new perspective on the cancer question, and perhaps move research in a more productive direction. Larry Nagahara, the NCI program director for this initiative, noted astutely that the questions a physicist would ask of cancer may help "shed light on how cancer develops as a disease." If all the king's horses and all the king's men could not put Humpty Dumpty back together again, then perhaps it was time to ask for help from somebody other than the king's men. The NCI funded twelve Physical Sciences–Oncology Centers with $15 million each to look into the question of cancer's origins and treatment.

Why was it such a big deal to bring in scientists from other fields to study cancer? Doctors and medical researchers follow the dictums of "evidence-based medicine." The status quo is considered to be fundamentally correct, and changing that understanding requires many peer-reviewed studies. Unfortunately, these studies often take decades and millions of dollars, so progress is glacial. Old paradigms of disease survive, even as patients die.

For example, in the 1960s, many people suspected that secondhand smoke caused the same diseases as smoking itself: lung disease and cancer. It was super obvious and logical. But without evidence from peer-reviewed studies, it was considered merely a hypothesis. So, decades of research and millions of dollars were needed to prove that secondhand smoke was indeed dangerous before some seemingly commonsense measures were enacted.

Smoking was not banned on airplanes until 1988. Smoking caused cancer, but for decades we allowed toxic, carcinogenic secondhand smoke to be circulated to all the passengers in an airplane. Restaurants had nonsmoking sections that, as if by magic, would protect patrons on the other side of the restaurant from the noxious smoke. That's how evidence-based medicine works: vigorously defending the status quo against new ideas. Each step along the way, to the truth is paved with decades of bickering and demands to "see the evidence." It would have been more sensible to force tobacco companies to prove that secondhand smoke was safe than to force medical researchers to prove that it was harmful. But because the status quo agreed that secondhand smoke was safe, it fell to the researchers to prove harm.

In nutrition, guidelines established in the 1970s advised Americans to severely restrict all dietary fat and to eat more carbohydrates. The original 1992 Food Guide Pyramid published by the U.S. Department of Agriculture (USDA) advised Americans to eat six to eleven servings every day from the bread, cereal, rice, and pasta group. The images used included pictures of such "healthy" foods as white bread, pasta, and crackers. The USDA also advised Americans to eat sparing amounts of foods such as avocados, salmon, nuts, and olive oil, due to misplaced fears about fat.

It would take decades before these whole, natural foods were considered acceptable and maybe even healthy—all because evidence-based medicine so vigorously defends the status quo. Even though the original dietary guidelines were drafted based on faulty scientific evidence, any change to them must be rigorously proven with multimillion-dollar studies. Since humans have been harmlessly eating avocados and olive oil for centuries, it would have been more sensible to force researchers to prove that traditional foods such as olive oil were harmful rather than prove they were safe.

Cancer medicine is no different. Once the genetic paradigm of the somatic mutation theory was established, it was considered sacrosanct. Even as damning evidence piled up against the SMT as a viable paradigm of cancer, researchers still clung to it like a drowning man clings to a life raft. Medical research insists that all new articles in scientific journals be reviewed by other scientists, who can demand changes before publication or refuse to publish outright. Radical new ideas are often rejected immediately, never seeing the light of day. Peer review is a search for consensus, which researchers presume is the truth. This ensures that old opinions hold and that new ideas are stifled.

Physics works differently. You may start with a classical theory like Newton's three Laws of Motion, but when you find an anomaly, such as the wave-particle duality of light, then you must come up with a different theory to explain it. Even if you cannot prove the existence of quanta of energy, if the new theory explains the known facts and the anomalous findings better than the original theory, then it supplants it. Thus, a Swiss patent clerk (Albert Einstein) was able to find support for his radical theories of general and special relativity long before there was actual proof. Whereas physics is always evaluating new theories, medicine is always looking to reject them.

Physics also embraces the anomaly, because science moves forward only by explaining anomalies. The great American physicist Richard Feynman once said, "The thing that doesn't fit is the thing that's the most interesting; the part that doesn't go according to what you expected." Medicine, on the other hand, rejects anomalies. If the consensus is that cancer is caused by genetic mutations, any anomalous data can be conveniently ignored.

The process of peer review does not tolerate dissension. New theories are published only if a number of other scientists agree. In physics, your theory is good only if it explains the known observations. In medicine, your theory is good only if everybody else likes it, too. This explains the rapid pace of progress in the physical sciences and the glacial pace of medical research. Medical research works well when it is already on fundamentally the right track, such as with infections, but it fails when dealing with a disease such as cancer, whose etiology is completely unknown. This may be due to the high consequences of a failed theory in medicine, where the price is paid in lives lost.

Physics moves forward in gigantic leaps and bounds—in quanta, if you will. A single correct theory, such as Einstein's relativity or Niels Bohr's quantum theory, moves the entire field an incredible distance. Medical science, by contrast, laboriously moves a single step at a time. This is how we spent decades vilifying all forms of dietary fat before spending millions of research dollars to show that some natural fats, like nuts and olive oil, are good for us after all.

Even within medicine, there are occasional breakthroughs. In heart disease, new procedures, technology (e.g., pacemakers), and medicines have slowly reduced the cardiovascular death rate over the last sixty years. Cancer? Not so much. The world of technology rides a bullet train; the world of medicine moves at a crawl; but cancer remains standing still. This despite billions of research dollars spent every year, more "walks for cancer" than you could walk, and more pink ribbons than you could count.

In 2014, the renowned oncologist Robert Weinberg noted that even by the 1970s, cancer research had generated an enormous amount of data, "but there were essentially no insights into how the disease begins and progresses to its life-threatening conclusions." As a result, Weinberg lamented that cancer research was held in "ill-disguised contempt" and that "one should never, ever confuse cancer research with science!"[7]

When he was called upon by the NCI, Dr. Paul Davies confessed that he possessed no prior knowledge of cancer. Good, said the NCI. That was exactly what they were looking for. Davies was interested primarily in astrobiology and had never really thought about cancer. This gave him the freedom to start with perhaps the two most basic questions: What is cancer? Why does it exist?

We had no satisfactory answers to these questions. What initiates the cancerous transformation of a cell? Why shouldn't *all* cells turn into cancer? Cancer cells originate and mutate from our own cells. But under what environment?

An even more profound question about cancer's origin had yet to be addressed: why can virtually every cell in the human body become cancerous? There are cancers of the lung, breast, stomach, colon, testicles, uterus, cervix, blood cells, heart, liver, and even fetuses. The ability to become cancerous is an *innate* capability of every cell in the body, almost without exception. Sure, some cells become cancer more frequently than others, but virtually no cell *cannot* become cancer. The oncogenes and tumor suppressor genes discovered so laboriously over the last quarter century are mutations of *normal* genes. *Every single cell* of our body contains the seed of cancer. Why?

This mystery goes deeper still. Cancer is not just a human disease. Davies noted that "What struck me from the outset is that something as pervasive and stubborn as cancer must be a deep part of the story of life itself. Sure enough, cancer is found in almost all multicellular organisms, suggesting its origins stretch back hundreds of millions of years."[8] Dogs get cancer. Cats get cancer. Rats get cancer. Even the most primitive multicellular organisms develop cancer. In 2014, cancer was discovered in two species of hydra. You may recall from high school biology class that hydra are simple, small aquatic organisms that evolved very early on from single-cell organisms.[9]

The origins of cancer are found at the origins of all multicellular life itself. This may have seemed obvious to a cancer outsider, but not to an insider with the curse of knowledge. "Cancer," Davies astutely noted, "is very deeply embedded into the way multicellular life is done."[10]

Cancer is older than humanity. Searching for the answers to cancer's origin among the evolutionarily recent genes of humans is futile. The answers are simply not there. Cancer was something much older and more fundamental to life on earth than humanity.

Most medical researchers and doctors view cancers as some kind of crazy genetic mistake. But to the outsider Davies, the behavior of cancer cells seemed anything but berserk. Instead, cancer appeared to be a highly organized, systematic technique for survival. It's no accident that cancer survives everything the body throws at it. It's no accident that cancer survives everything modern medicine throws at it. It survives chemotherapy, the most devastating poisons in our pharmacopeia. It survives radiation. It survives our best surgical attempts. We've spent decades developing humanized antibodies with the most precise genetic weaponry ever seen—and still, cancer only laughs. That's not random. It's highly organized. We thought cancer was crazy, like the Joker, when it is actually more akin to Lex Luthor: wickedly smart.

Cancer must develop and coordinate many "superpowers" that allow it to survive. It grows. It is immortal. It moves around. It uses the Warburg effect. Did all these miraculous attributes precisely come together at the exact right place and the exact right time just as a freak accident?

That is as likely as throwing a pile of bricks into the air and having them land exactly in the form of a house. Further, how can such a freak accident happen to every cell in the body, in every multicellular organism known to exist? If something seems "stupid" but works (survives), then by its very definition, it's not stupid. Yet we had viewed cancer as some kind of random collection of stupid genetic mistakes. Yes, there was stupidity going on, but it wasn't the cancer's.

The "preposterous reductionism" of looking only at the genetics of cancer had failed. We missed the forest for the trees, but a new paradigm was taking hold that would bring fresh new insights into the very origins of cancer.

11

THE ORIGINS OF LIFE AND THE ORIGINS OF CANCER

ACCUSTOMED TO THINKING about life on other planets, cosmologist Paul Davies wondered how cancer fit into the story of life on Earth. Because cancer is as old as multicellular life itself, the origins of cancer, he reasoned, must lie in the origins of life.

So, let's take a step back. How did life on Earth evolve?

Life on Earth began an estimated 3.8 billion years ago, perhaps 750 million years after Earth's formation.[1] Simple organic molecules may have formed spontaneously in Earth's early atmosphere. Stanley Miller's famous experiments in the 1950s showed that electrical discharges into a mixture of hydrogen, ammonia, and water that replicated the early atmosphere could produce simple amino acids. But these organic molecules were not yet cells.

The earliest cells were created when self-replicating molecules called ribonucleic acids (RNA) were enveloped in a membrane called a phospholipid bilayer, which is still the basis for all modern human cell membranes. This bilayer protected the RNA from the harsh outside environment, allowing self-replication. These early cells lived in a sea of nutrients, obtaining food and energy directly from their environment. As long as nutrients were available, they survived, but they were always on the edge of extinction.

The prime directive of life, even at this early stage of evolution, was to replicate. Reproduction demands growth, cellular energy generation, and the ability to move around to find more favorable environments. Even viruses, nonsentient pieces of nucleic acids that straddle the border of the definition of life, have a biological imperative to replicate. They may not be fully alive, but they are programmed to replicate and require a host cell's help to do so.

Prokaryotes are the earliest and simplest organisms that evolved from the primordial soup. It took another 1 billion to 1.5 billion years to evolve the more complex eukaryotes that contained organizing features like a nucleus and organelles. The specialized nucleus carried all the genes necessary for reproduction. Organelles (literally, miniature organs) are subcellular structures that allow the compartmentalization necessary for specific functions such as protein production and energy generation.

The organelle called the mitochondrion generates energy for the cell. Unlike other organelles, mitochondria are believed to have originated as separate prokaryotic cells. As early eukaryotic cells became more complex, mitochondria discovered they could live inside these cells in a mutually beneficial relationship. Mitochondria were protected inside the cell, and in return, they generated energy in the form of adenosine triphosphate (ATP). This relationship evolved over time, and today, one cannot survive without the other. Mitochondria are present in all mammalian cells except for red blood cells.

The mitochondria contain their own distinct DNA, which reflects their origins as separate cells. Although the generation of ATP by oxidative phosphorylation (OxPhos) is considered their main function, mitochondria are also key regulators of apoptosis, a method of controlled cell death.

Early in planet Earth's history, in the Proterozoic age, the atmosphere was largely devoid of oxygen, and most cells generated energy anaerobically (without oxygen). Earth's atmosphere began to change with the rise of photosynthetic organisms. Energy from sunlight combined with carbon dioxide to give off oxygen as a waste product, which slowly accumulated in the atmosphere.

This was a big problem for the other early cells, as oxygen is toxic if not handled properly. Our body includes robust antioxidant defenses for precisely this reason. The mitochondria use this oxygen to their advantage by metabolizing glucose through OxPhos. This generated ATP more efficiently but also neutralized some of this toxic oxygen. As a result, present-day mammalian cells have functional pathways for both aerobic (OxPhos) and anaerobic (glycolysis) energy production, and the ratio can vary depending upon energy needs.

The transition from simple prokaryotic cells to the more complex eukaryotic cells, complete with specialized organelles and mitochondria, was a huge evolutionary jump. Protozoans (e.g., yeast) are simple, single-cell eukaryotic cells, but they are vastly more complex and larger than bacteria. All living creatures for the first half of the history of life on Earth were single-cell organisms. The next big evolutionary hurdle was multicellularity.

THE JUMP TO MULTICELLULARITY

Single-cell organisms are selfish creatures; they live, grow, breed, and pretty much do everything else by themselves. There is nobody for them to help out and nobody to help them. Their prime directive is their own survival and reproduction. To be successful, a single-cell organism competes with surrounding cells for resources. But cells working together have a huge advantage over cells working alone.

Multicell organisms evolved about 1.7 billion years ago, likely beginning as simple aggregates or colonies of single-cell eukaryotes. Over time, mutually beneficial collaboration between cells permitted specialization, which then led to true multicellular organisms. Specialization, division of labor, and intercellular communication made these organisms bigger, more complex, and more capable than simpler single-cell organisms. The human body contains more than two hundred types of these specialized cells, which are broadly classified into five categories: epithelial tissue, connective tissue, blood, nervous tissue, and muscle.

But this new complexity demanded new rules of multicellular cooperation. When grouped together, individual cells must learn to live and work together just as individual people in large cities do. A single-cell organism is like an individual living alone in the woods. He may do whatever he likes; nobody else is around to care. He can walk around naked all day if he wants. Multicell organisms are like large, densely populated cities. There must be rules to govern acceptable behaviors. A man walking around naked may be arrested. The needs of the many outweigh the needs of the individual. In return for the sacrifice of a few individual freedoms, societies allow specialization, division of labor, and communication. This increased complexity allows cities and nations to dominate their environment.

A multiperson city or a multicell organism prioritizes decisions to benefit the collective. In a city, some individuals die so that others benefit, such as soldiers, firefighters, and policemen. In a multicell organism, cells such as the white blood cells of the immune system may be sacrificed for the good of the many cells within the entire organism.

Cells must follow strict rules for cooperation and coordination if they are to live and work together. The priority for single-cell versus multicell organisms changes significantly. Single-cell organisms *compete* with other cells to benefit themselves. Multicell organisms *cooperate* with other cells to benefit the entire collective of cells that make up the organism.

Multicell organisms compete with other organisms for food, but at a cellular level, all the cells within that organism cooperate (see Figure 11.1).

	Single-Celled Organism Single Person	Multi-celled Organism Multi-Person City
Priority	Individual	Entire organism/city
Modus Operandi	Competition	Cooperation

Figure 11.1

These cell-level differences between single-cell and multicell organisms manifest in several important ways: growth, immortality, movement, and glycolysis.

Growth

Single-cell organisms grow and replicate at all costs. That is their entire purpose in life, and their default state. A bacterium in a Petri dish or yeast in a slice of bread never stops trying to grow and reproduce. It does not stop until resources run out.

By contrast, multicellular organisms impose tight control over growth using genes that promote growth (oncogenes) and genes that suppress it (tumor suppressor genes). Cells may grow only when they are told—in the right place and at the right time. A liver cell cannot grow on the tip of your nose. Also, liver cells cannot grow to the size of a refrigerator; this would impact the lung, living right next door. Good fences make good neighbors. This ensures the well-being of the entire *organism*, not the individual *cell*.

Similarly, the single person and the multiperson city differ substantially in their approach to growth. The lone survivalist in the woods faces no restrictions on growth. He may build his house as large as he wants, and wherever he wants. Growth is generally good. Cities, by contrast, impose tight control over growth. You cannot simply build a shed on your neighbor's property. There are rules that ensure cooperation. Growth is generally bad because there is limited space available. If you grow, it will be at your neighbor's expense. Growth of the whole city is good, but growth of the people within that city is bad if the city itself is not expanding.

Immortality

Single-cell organisms are immortal because they can replicate infinitely. There is no limit to how many times a single-cell organism like yeast can divide. For example, there are sourdough yeast starters that are more than a hundred years old that are still used to make bread.[2] The yeast grows and replicates indefinitely as long as conditions are right. The yeast line is immortal.

Cell lines in a multicellular organism are not allowed to live forever. Each time they replicate, their telomeres get a little shorter, and when they are at a critical length, the cells can no longer divide. At that point, the cell line has reached senescence. Decrepit cells that have divided too many times are condemned to die through apoptosis. Once they have outlived their useful lives, they are removed for the good of the organism.

A lone survivalist in the woods may keep his house as long as he wants, even if the roof leaks and the walls are about to fall down. In a city, when houses get too old, they are condemned and destroyed so that other people do not get hurt. The needs of the many take precedence over the needs of the individual.

Movement

Movement is the natural state of single-cell organisms. They have no particular obligation to stay in any specific place. They move around to find the most favorable environment. Bacteria have evolved to move in many spectacular ways. Some bacteria use an organelle called a flagellum, a long structure that acts much like a propeller. Other bacteria use twitching and gliding movements made possible by an organelle called a Type IV pilus.

Single-cell organisms also take advantage of passive movement. For example, when conditions are unfavorable, yeast enters a dormant state called a spore, which can be picked up and scattered by the wind. Some will find a favorable growth environment, reactivate, and bloom. Others will not, and will continue to lie dormant. Baker's yeast, for example, may stay in the little plastic packet for years and still reactivate when placed in warm water.

Movement is particularly advantageous to single-cell organisms' survival because they depend so heavily on their environment to provide for their needs. Yeast that stays in the same location for too long may exhaust its resources and perish. Being able to move means it can find more abundant resources elsewhere to thrive and reproduce.

By contrast, multicellular organisms must ensure that their cells remain anchored to their proper location and *don't* move around. Cells interact and depend on one another, so they must be in the right place at the right time. The liver depends upon the lung cell to gather oxygen, and the rest of the body depends on the liver to detoxify the blood. For this to work, everybody must be in the right position. The lung cell can't just hop into the bloodstream and take a ride downtown to hang out with the liver. Multicellular organisms have evolved complex systems called adhesion molecules to attach cells to their proper position.

The default state of single-cell organisms is movement, and the default state of cells with a multicell organism is staying put (stasis). Movement occurs at the level of the entire organism, not at the level of the individual cell. Organisms move around, but the cells within that organism do not.

A man living alone in the wilderness may move around wherever he wants. If conditions are good in one spot, he may stay. If not, he may move to a better location. Early human tribes were often nomadic, roaming the countryside looking for food and evading enemies. But a man living in New York City cannot simply move wherever he wants. He can't just walk into another person's house. This is one of the many rules of living in a society.

Glycolysis

Energy generation evolved in three stages: glycolysis, photosynthesis, and oxidative metabolism.

Earth's early atmosphere was largely devoid of oxygen (anaerobic conditions), and thus the earliest evolved form of energy generation was glycolysis. This process breaks down a glucose molecule for two ATP and two lactic acid molecules and does not require oxygen. All modern human cells have the capacity to undergo glycolysis.

The next major evolutionary step in energy conversion was photosynthesis, which arose approximately three billion years ago. Proliferation of photosynthetic bacteria caused rising atmospheric accumulation of oxygen.

The increased availability of oxygen set the stage for the evolution of the third major type of energy generation: oxidative phosphorylation, or OxPhos, using the mitochondrion. OxPhos burns glucose with oxygen to provide thirty-six ATP per glucose, a massive upgrade from the two ATP produced by glycolysis. OxPhos is almost universally used in modern human cells when oxygen is available. While most single-cell organisms use the more primitive glycolysis, most eukaryotic cells use OxPhos.

So, to summarize, single-cell organisms differ from multicell organisms by these following four main characteristics:

1. They grow.
2. They are immortal.
3. They move around.
4. They use glycolysis (also called the Warburg effect).

Does this list look familiar? It should; this is precisely the same list of attributes that make up the four hallmarks of cancer! (See Figure 11.2.) Surely this is not a coincidence. The hallmarks of cancer are also the hallmarks of unicellularity. Cancers are derived from cells that are part of a multicellular organism, but their behavior closely resembles that of a single-cell organism.

Hallmarks of Cancer	Unicellular	Multicellular
Grow	Yes	No
Immortal	Yes	No
Move	Yes	No
Glycolysis (Warburg)	Yes	No

Figure 11.2

Cancer cells differ from normal cells precisely in the way that single-cell organisms differ from cells within a multicell organism. It's like the answer to a college SAT question: cancer cells are to normal cells as single-cell organisms are to cells in a multicell organism. Considered from this vantage, we can see even more similarities between cancer cells and single-cell organisms.

SPECIALIZATION

A person living alone in the forest must perform all the tasks of survival: gathering food, hunting, protecting himself, sewing clothes, etc. He does not live long if his only skill is performing tax audits. A society enables people to specialize: farmers, hunters, bakers, merchants, etc. Cooperation and coordination permit greater efficiency, and this increased complexity eventually allowed humans to reach outer space, build supercomputers, and conquer the atom. But these benefits of specialization come at the cost of other functions.

Single-cell organisms can rely on only themselves to perform all the functions necessary for life, and therefore cannot specialize to perform a single function. Microscopic descriptions of cancer cells characterize them as primitive or dedifferentiated (less specialized). As cancer progresses, cells become more primitive in appearance, with progressive loss of "higher" specialized functions. The term *anaplasia*—derived from the Greek *ana*, meaning "backward," and *plasis*, meaning "formation"—is often applied to cancerous cells. Cancer cells seem to be moving backward in evolution.

This is most obvious in blood cancers such as acute myelogenous leukemia (AML). Normal bone marrow produces immature white and red blood cells called blasts. When mature, they are released into the bloodstream. These blasts normally constitute less than 5 percent of the bone marrow and are not found in the bloodstream. AML is defined by the presence of more than 20 percent of these immature blast cells in the bone marrow. They are often also found in the bloodstream, an ominous sign. Cancerous progression is the move toward less developed, more primitive, and less specialized cellular forms.

Cancer shifts away from specialized function and toward pure reproduction and growth. Normal breast cells are specialized to make milk when needed. A breast cancer cell, for its part, is not primarily concerned with milk production, but with growth of more breast cancer cells. A colon cancer cell no longer concerns itself with absorption of nutrients, but is concerned mainly with its own growth and replication.

By contrast, multicellularity allows division of labor and specialization in structure and function. This increased size and complexity allow it to dominate its environment. Liver cells are specialized to function with much higher efficiency. But they become so specialized that they cannot survive alone. You can put some bacteria on the ground, and they may flourish. But put a piece of liver onto the ground, and it will certainly die.

AUTONOMY

The lone survivalist in the woods has complete autonomy. The man who lives in New York City must follow many rules and laws.

He must pay his taxes. He must follow his condominium's code of conduct. He must adhere to societal norms.

Single-cell organisms are their own boss, with complete autonomy. Cancer cells are the same; they do not follow the rules. Breast cancer cells will not respect the borders of the breast, but will metastasize to other organs. Breast cancer cells do not respond to orders from the brain or hormones or any of the other normal controlling methods the body uses. Breast cancer cells grow for their own good, not the good of the organism.

In multicellular organisms, individual cells must do exactly as they are told. Hormones carry detailed instructions on what to do. If the hormone insulin is high, then cells cannot refuse entry to glucose. They have no autonomy. Cells have no existence outside the whole organism. Your lung doesn't rummage around in the refrigerator at night. We don't stop to say hi to our neighbor's liver while out walking the dog. You don't yell at your kidney to put the toilet seat down.

HOST DESTRUCTION

The lone survivalist may or may not care for his surrounding environment. He may dump garbage in the river, to be carried away and become somebody else's problem. A city, however, carefully regulates the local environment. Garbage must be deposited in certain places. You do not drive on your neighbor's carefully manicured lawn.

Single-cell organisms take no responsibility for the environment around them. A yeast will do whatever it can to kill its bacterial neighbors, because they are competition for food and other resources. Sir Alexander Fleming observed the penicillium mold secrete a substance that killed all surrounding bacteria. This led to the discovery of the world's first modern antibiotic, penicillin.

Cancer cells, like unicellular organisms, are locally destructive. A cancer will grow at the expense of its neighbors, destroying any surrounding tissue. The worse it is for its neighbors, the better it probably is for the cancer. Cancer is the guy who deliberately drives his pickup truck over his neighbor's lawn. Competition can involve making yourself better or making your competitors worse. Both strategies work. Welcome to the jungle.

As in a society, cells in a multicellular organism must be good neighbors. Multicellular organisms must maintain the extracellular environment (called the extracellular matrix) so as not to harm their neighbors. Normal liver cells, for example, cannot simply dump their waste next door into the lung's backyard. Normal breast cells can't start destroying neighboring skin cells.

EXPONENTIAL GROWTH

Single-cell organisms grow by dividing into two daughter cells. With sufficient resources, the population doubles with each generation, resulting in very rapid exponential growth. This exponential growth is typical of cancer but not of cells in multicellular animals. The adult liver, for example stays roughly the same size because the millions of new liver cells being created are balanced by an equal number of dying cells. As noted previously, multicell organisms maintain tight control over growth, not allowing unbridled population expansion.

INVASION INTO NOVEL ENVIRONMENTS

Single-cell organisms often invade and exploit new environments in their unending search for more food. Yeast mold growing on a slice of bread will continue spreading until it covers the entire slice.

Cancer, like single-cell organisms, invades everywhere and can colonize new environments in the process of metastasis. Breast cancer cells can survive in the liver. Lung cancer cells can survive in the brain. Infections, too, are often said to metastasize. An infection may start in the kidneys, spread through the bloodstream, and infect the heart valves. These metastatic infections are commonly lethal, too.

Cells within multicellular organisms maintain clear boundaries; they cannot survive outside their designated areas. A normal breast cell cannot survive in the liver, a completely foreign environment. A lung cell cannot survive in the brain.

COMPETITION FOR RESOURCES

Single-cell organisms compete vigorously for resources. It's every bacterium for itself. Cells that grab enough food will survive to reproduce. Those that do not will die. Cancer similarly competes for resources directly, with no thought for the ultimate good of anybody else. A cancer cell will use all the glucose it can, even if it must deprive normal cells. Cancer patients often lose extreme amounts of muscle and fat, as the cancer gorges itself. This process, common in most advanced cancers, is called cancer cachexia.

Cells in multicellular organisms do not directly compete with one another for resources such as glucose. When resources are scarce, there are clear rules for division. For example, during times of starvation, menstruation and reproductive ability are suspended, hair production slows, and fingernails become brittle. Scarce resources are directed toward survival of the organism, and some individual cells may be sacrificed. Relatively superfluous cells undergo apoptosis.

GENOMIC INSTABILITY

Genetic variation allows a species to evolve and survive in unpredictable environments. Single-cell organisms reproduce asexually, splitting into two daughter cells that are genetically identical to the parent. If the genes are reproduced with 100 percent fidelity, there will be no genetic variation whatsoever. To create genetic diversity, single-cell organisms must mutate.

Microorganisms often elevate their rate of genetic mutation in response to stress using complex mechanisms such as aneuploidy,[3] slipped-strand mispairing, polymerase slippage, gene amplification, deregulation of mismatch repair, and recombination between imprecise homologies.[4] These processes sound complex because they are. The point is that necessity is the mother of invention: single-cell organisms find ways to increase mutation rates when needed.

Cancer, as has been painstakingly noted, is also full of genetic mutations. Cancer can mutate its genes better than almost anything else in existence. Gene mutation is one of the hallmarks of cancer, a foundational ability that makes cancer . . . well, cancer. For single-cell organisms, and cancer cells, the ability to mutate is a good thing; for multicellular organisms, it is a bad thing.

Multicellular organisms produce genetic variation by reproducing sexually, which mixes parental genes, but even when different sets of genes are combined, genomic stability is favored. Cells are so highly interdependent that a mutation in one cell will usually adversely affect another. If a lung cell mutates and no longer functions, it will adversely affect the rest of the body. A mutation in one hormonal pathway will likely impair another and have a domino effect. Thus, multicellular organisms evolved DNA-repair mechanisms to slow this natural mutation rate.

Mutations allow unicellular organisms to develop genetic varia-

At the Cellular Level	Multi-Celled Organism	Cancer Cells	Single-Celled Organism
Priority	Organism	Cell	Cell
Modus Operandi	Cooperation	Competition	Competition
Growth	No	Yes	Yes
Immortality	No	Yes	Yes
Movement	No	Yes	Yes
Glycolysis	No	Yes	Yes
De-Specialization	No	Yes	Yes
Autonomy	No	Yes	Yes
Host Destruction	No	Yes	Yes
Exponential Growth	No	Yes	Yes
Invasion/ Novel Environments	No	Yes	Yes
Competition for Resources	No	Yes	Yes
Genomic Instability	No	Yes	Yes

Figure 11.3

tion to deal with environmental instability. Cells in a multicellular organism don't need to deal with environmental instability, because conditions are held relatively static. The ionic composition of the surrounding fluid is held within very tight limits. Body temperature is relatively constant (see Figure 11.3).

Trait	Infection	Cancer	Heart Disease
Invade tissue?	Yes	Yes	No
Metastasize?	Yes	Yes	No
Develop resistance?	Yes	Yes	No
Develop genetic mutations?	Yes	Yes	No
Evolution of cells?	Yes	Yes	No
Secretion?	Yes	Yes	No

Figure 11.4

This paradigm of cancer as an invasive protozoan explains why cancer resembles an infection much more closely than other human diseases such as heart disease.

THE EVOLUTIONARY PARADIGM

Cancer originates from cells of a multicell organism, but it behaves precisely as a single-cell organism. This is a spectacular and novel finding. At long last, we have a new answer to the age-old question: what is cancer? The conventional answer from cancer paradigm 2.0 had long been that cancer is a cell with randomly accumulated genetic mutations. But Davies and others saw that the origins of cancer lie in the origins of life itself. Cancer is, improbably, a single-cell organism. Multicellular life is about cooperation. Unicellular life is about competition (see Figure 11.4). This type of throwback to an earlier ancestral phenotype is called an atavism, the default to an earlier version, or the return to an evolutionary past.

Human civilization has evolved from small groups of individuals competing with one another to large societies working together. This increased size, complexity, and specialization allow

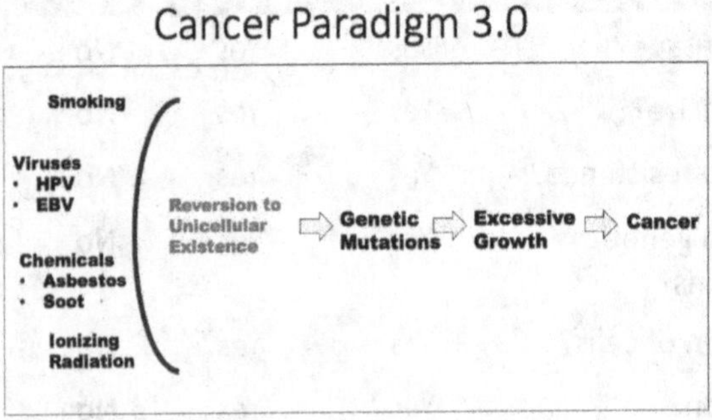

Figure 11.5

cities to dominate. Similarly, life on earth has evolved from unicellularity to multicellularity. The increased size, complexity, and specialization allow multicell organisms (such as humans) to dominate (see Figure 11.5). Cancer is like the postapocalyptic world of *Mad Max*, where small bands of people fight one another for resources.

The city dweller and the lone survivalist in the woods may appear to be completely different, but really, they are similar, just facing different situations. In the woods, people compete. In the city, people cooperate. But what happens in a city when law and order break down? The city dweller acts more and more like the survivalist. The problem is not only the seed; it's also the soil.

Cancer is the breakdown of multicellular cooperation. Cancerous transformation happens when a cell in a well-functioning society acts like a single-cell organism. Just as a city has laws in place, normal cells have strong anticancer mechanisms, which include the cells of the immune system. When these are overwhelmed and the rules of cellular cooperation break down, cells must revert to their original programming. As it stops following the rules, cancer will prioritize only its own survival.

Without cooperation, you compete or you die. This reversion toward unicellularity has devastating results for the organism. Because all multicellular life evolved from unicellular organisms, all multicellular life contains the basic pathways needed for cancer. The seeds of cancer are therefore contained within *every cell of every multicellular animal*. The origins of cancer lie in the origins of multicellular life on earth itself.

But how did that cell, originally part of the multicellular community, change its behavior to that of a single-cell organism? Only one force in the biological universe has that power.

Evolution.

12

TUMORAL EVOLUTION

CHARLES DARWIN, STUDYING animals in the idyllic Galápagos, documented his history-making theory of evolution by natural selection in his 1859 book *On the Origin of Species*. According to legend, Darwin noticed that the shape and size of the beak of what he presumed to be a finch—it may actually have been a tanager—varied according to the bird's food source. Some birds had long, pointy beaks that were great for eating fruit, and other birds had shorter, thicker beaks that were great for eating seeds off the ground. Darwin reasoned that it could not be simply a coincidence that the pointy-beaked birds were common where fruits were plentiful and the thicker-beaked birds were common where seeds were plentiful.

He considered another bird, which he identified as the domesticated pigeon but which may actually have been a rock dove. In the 1800s, pigeon fanciers bred these birds for specific attributes. A pigeon breeder who wanted a white pigeon would breed together only those pigeons with very light coloration over many generations. Eventually, the result was a pure white pigeon. If a pigeon breeder wanted a bird with huge feathers, he selectively bred together birds with the biggest feathers over many generations. Artificial selection eventually produced a bird with the desired feature. It is not necessary to know the specific gene mutations desired, only the criteria for selection.

Artificial selection has been used for thousands of years. To produce dairy cows, the most prolific milk-producing cows were bred together over many generations. The other cows became beef stew. Eventually, you got a Holstein cow (the one with the familiar black-and-white-spotted pattern), which can produce upward of thirty liters of milk per day. Different genetic mutations can produce the same result. The Brown Swiss cow is also a great dairy cow but has completely different genetics from Holstein's. These genetic variants are not random mutations, but were created with a specific target in mind: milk production.

Darwin reasoned that the same selection process was happening in the Galápagos finches. Instead of artificial, man-made selection for specific traits, he postulated that there was a natural selection process. Areas with plenty of fruit favored the survival of birds with longer beaks. When bred together, these birds with longer beaks produced more long-beaked birds.

Long, pointy beaks were not the result of random genetic mutation (the seed), but of the environmental condition (the soil): plenty of fruit allowed for pointy-beak mutations, out of all possible beak configurations, to flourish. The opposite happens if the main food source is seeds—in which case, short beaks confer an advantage to the birds.

Changes in a population through selection, whether artificial or natural, have two prerequisites: genetic diversity and selective pressure. If every bird had the same beak or every cow produced the same amount of milk, evolution through natural selection would not be possible, because all choices would be identical. There is no natural advantage or disadvantage. Natural selection explains the process by which certain traits appear or disappear. The environment exerts selective pressure to determine which genetic changes are most favorable for survival. The soil determines which seeds will flourish. The somatic mutation theory suggested that cancer cells were genetically monotonous and that mutations accumulated randomly, rather than through any selection process. These assumptions could not have been more incorrect.

INTRATUMORAL HETEROGENEITY

Does cancer contain the necessary genetic diversity to allow evolution? The answer is a resounding yes, as shown by the Cancer Genome Atlas. Significant genetic variation exists even within a single tumor mass; this is called intratumoral heterogeneity (ITH). The prefix *intra* means "within," and *heterogeneity* means "the state of being diverse," so ITH refers to the startling diversity of genetic mutations found within a single cancer mass, or tumor.[1]

Tumors that shared many similar characteristics (hallmarks) differed tremendously when seen on a genetic scale.[2] Even within the same patient, different sites of a single tumor exhibited vastly differing genetic mutations.[3] For example, in a 2012 study, researchers biopsied a single cancer from a single patient. Nine samples from the primary tumor and three from various metastatic sites in the same patient's body were genetically sequenced and compared. Where the SMT had predicted 100 percent genetic concordance, the truth was quite different: only 37 percent of somatic mutations were shared. Cancer is not a single genetic clone; instead, it contains multiple diverse subclones.

Most cancers harbor a dominant clone that comprises more than 50 percent of the tumor, with the remainder containing multiple genetically diverse subclonal populations. Sometimes there are incredible genetic differences between cancer cells within a single tumor. In one case study, the dominant clone differed from a subclone by 15,600 genetic mutations![4]

Tumors are genetically diverse through space, but also over time.[5] New mutations are constantly springing up, while other mutations are dying. One study compared the genome of a recurrent metastatic breast cancer to its original genome taken nine years before. There were nineteen new mutations in the metastasis that were not present in the original tumor.[6] The genetic diversity of ITH is a key enabler of tumor evolution, allowing natural selection to proceed through branched-chain evolution.

BRANCHED-CHAIN EVOLUTION

How does tumor evolution proceed? The SMT proposed that cancer evolved linearly. Cancer cells added mutations one at a time, until the cells acquired all the hallmarks needed to become cancer. This theory predicted that a single disruption, such as a drug or engineered antibody, could break the entire chain and cure cancer. A fantastic story, but now known to be wrong for most common cancers.

Instead of linear evolution, ITH permits the more robust process of branched-chain evolution. Cancer does not evolve along a single chain but along multiple tracks, much like a tree grows by sending out multiple branches. An obstacle to one branch does not block the tree's overall development because other branches more advantageously situated will continue to grow (see Figure 12.1).

Consider a tree growing near a fence. If the tree had only a single branch, it would stop growing upon hitting the wooden fence.

Figure 12.1

But because the tree has multiple branches, it grows through the fence almost unimpeded by finding and exploiting openings in it. Most species evolve in a similar manner. For example, Darwin's finches had every sort of beak; in some situations the long beak is favored, and in other situations the short beak is favored.

Cancer is also now known to follow branched-chain evolution. Figure 12.2 illustrates how ITH and branched-chain evolution allow greater survival. When cancer encounters an obstacle—for example, the administration of chemotherapy that kills 99 percent of cancer cells—only a single subclone of cancer need survive in order to repopulate the tumor and allow the evolutionary process to continue. A tree with multiple branches needs only a single hole to grow through a fence.

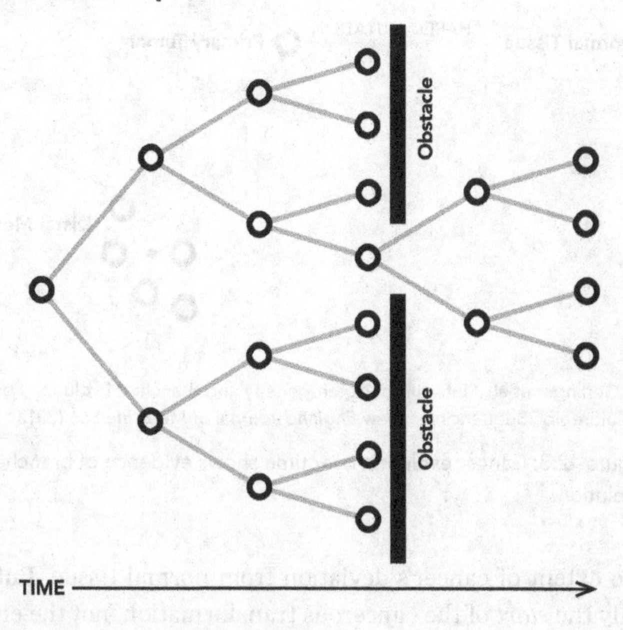

Figure 12.2: Genetic heterogeneity and branched-chain evolution of cancer cells allow cancer to adapt to obstacles.

Recent research has been able to follow the evolutionary changes in a cancer patient. Figure 12.3 illustrates how it is possible to map the genetic mutations of a single cancer and its evolution over time. The specific mutations are not as important as noting how cancer mutations evolve like the branches of a tree.

From normal tissue, all cancer cells begin with a single common mutation. The SMT assumed that this single mutation was

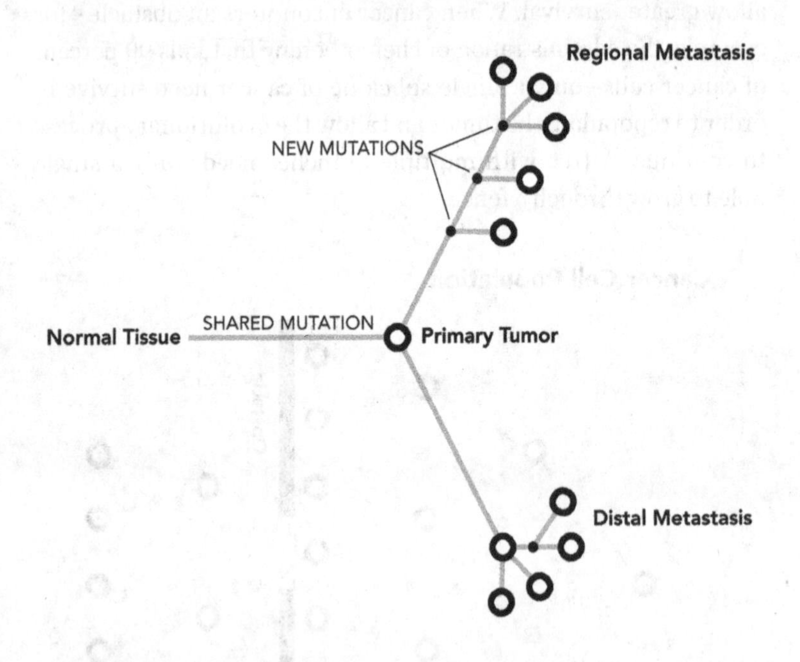

M. Gerlinger et al., "Intratumor Heterogeneity and Branched Evolution Revealed by Multiregion Sequencing," *New England Journal of Medicine* 366 (2012): 883–92.

Figure 12.3: Cancer evolution over time shows evidence of branched-chain evolution.

the extent of cancer's deviation from normal tissue. But it was only the *start* of the cancerous transformation, not the end. New cancer mutations branched off from the main stem, and over time, more and more new branches evolved.[7] When the cancer encounters a problem that impedes its growth, one of its many different subclones may provide a solution. That subclone may continue proliferating to become dominant, and the overall tumor continues its growth. The obstacle, then, acts as selective pressure.

THERAPEUTIC IMPLICATION

The recognition that cancer continually evolves through both time and space via branched-chain evolution was a major break from the preceding decades of cancer orthodoxy. This has two major implications for cancer treatment and explains much of the lack of progress in oncology.

1. A single targeted treatment is unlikely to be successful.
2. Cancers can evolve to be resistant to treatment.

First, most cancers share only a minority of genetic mutations throughout their genome. Therefore, a single drug that targets a single mutation is unlikely to successfully treat the entire tumor. The dream of targeted, personalized medicine to shut down the one or two mutations of cancer was now well and truly dead.

There were exceptions, to be sure. Targeted medicine worked spectacularly well in chronic myelogenous leukemia and in *HER2/neu*-positive breast cancer. But with most cancers containing hundreds of mutations, this strategy could not work. Dozens of different drugs would be required for each genetically distinct site of cancer, including the metastatic sites.

Think about a tree. You can cut it down with a single ax blow aimed at the trunk, but it's difficult to chop through it. If you remove side branches, you won't likely impede the tree's overall growth, because you are pruning the tree rather than cutting it down. Cancer is the same. The trunk is usually too tough to cut, and targeting hundreds of small side branches is inefficient.

The logistics of targeting multiple mutations is daunting. A single biopsy of a tumor will miss the majority of genetic abnormalities. Even if you knew all the mutations present, it would require tens or hundreds of drugs in combination to affect all the branches. The idea of "precision" chemotherapy was based on an erroneous assumption of linear evolution. Whereas heterogeneity is dynamic and evolves over time and space, our treatments are static.

The research produced by the fifty-year-old war on cancer had catalogued millions of possible ways that genes mutated, with the misguided hope that this information would lead us to a cure. It has not come close. Targeting a single mutation when cancer has hundreds of them is not a fruitful strategy.

The most widely mutated gene in human cancers, defective in approximately 50 percent of them,[8] is called *p53*, and it was discovered in 1979. It is sometimes called the "guardian of the genome" because of its importance in maintaining genomic stability. Any form of DNA damage, as might be caused by toxins, viruses, or radiation, activates the *p53* gene. If the damage is minimal, *p53* will simply repair the damaged DNA. However, if the damage is too severe, then *p53* hits the kill switch and activates the apoptosis protocol, thus guarding the genome against defective cells.

Since *p53*'s discovery, approximately 65,000 scientific papers have been published about this gene alone. At a conservative cost of $100,000 per paper—incidentally, this is likely way, way too low—this research effort cost $6.5 billion. That's "billion" with a capital *B*. Since 1979, an estimated seventy-five million people have developed *p53*-related cancers. What do we have to show for this enormous cost, both in dollars and human suffering? The grand total of *p53*-based treatments approved by the FDA in 2019 is zero. Yes, zero. Why is it so hard to find a good treatment? So far, 18,000 different mutations of this gene have been identified.

In addition, branched-chain evolution allows cancer to develop drug resistance, a phenomenon commonly seen in bacterial infections. A genetically diverse population of bacteria adapts to antibiotic use by developing resistance. The first time an antibiotic is used, most bacteria are killed. Eventually, a rare mutation allows one bacterium to live. It thrives because the other bacteria have died, and there is no competition. So, another infection begins, but this time, the bacteria are resistant to antibiotics. Cancer behaves like an invasive protozoan, often becoming resistant to chemotherapy, radiation, and hormonal-based—and even the newer gene-based—treatments.

ITH and branched-chain evolution are potent survival mechanisms found in almost all life on earth. Together, they allow adaptation to new environments. This explains the extraordinarily high failure rate in drug development; cancer drugs fail almost three times as often compared to drugs developed to combat other diseases.[9]

Branched-chain evolution provides a conceptual framework for understanding cancer therapy. To be successful, you must go big or go home. Going big by attacking the "trunk" mutations is likely to be effective but difficult. On rare occasions, we are able to find a treatment that can cut down the trunk, such as imatinib or trastuzumab. Here, a single snip of the chain is all that is needed to reverse the disease.

Another successful strategy is to overwhelm the cancer with multiple different treatments. This involves the simultaneous use of multiple chemotherapies (drugs) and multiple modalities such as surgery and radiation. And sometimes this treatment works well. Many leukemias and other childhood cancers are cured by combination chemotherapy. One of the earliest breakthroughs in chemotherapy was the combination of multiple agents into a single regimen. Today, few chemotherapy drugs are given alone. Instead, three or four drugs are given together in precise regimens.

This is the same strategy used for some infections. Because both infections and cancers behave like single-cell organisms, the similarities are not coincidences. Tuberculosis, caused by a slow-growing bacterium, requires several antibiotics given simultaneously. If you can kill 100 percent of the microbes, then resistance has no chance to develop.

The genetically nimble cancer is constantly evolving. A static gene-targeted treatment is easily outmaneuvered. Cancer plays chess, an ever-changing, constantly evolving, strategic game. Using a single targeted genetic treatment is like relying upon a single set move at a single set time. It will almost always fail.

Cancer had always been considered a single genetic clone, so evolutionary processes were considered irrelevant. But the realization that cancers evolve was electrifying. For the first time in decades, we had a new understanding of how cancer develops. The entire field of science known as evolutionary biology could now be applied to understand and explain why cancer develops mutations.

Cancers are constantly evolving, meaning that they are moving, not static, targets. The key to hitting a moving target is knowing the driving forces behind it, so the key to hitting cancer is knowing the driving forces behind those mutations. What are those selective pressures?

SELECTIVE PRESSURE

Cancer's deep evolutionary roots run way past the origins of mankind, right back to the origin of multicellular life on earth. So, what is cancer? The simple answer is this: Cancer is a unicellular organism, but in order to transform itself from a normal cell in a "society" with rules for cooperation to a unicellular existence, it had to undergo hundreds or thousands of genetic mutations. The next question to answer is: what guided the selection of these mutations?

The somatic mutation theory assumed that cancer was simply

a collection of random genetic mistakes. But cancerous transformation is clearly not random. Instead, cells evolve toward a clearly defined destination, unicellularity, with the resolute purpose and tenacity of a bloodhound. Cancer cannot exist outside the host and is not transmissible; a highly successful cancer kills its own host and ultimately kills itself in the process. The more lethal the cancer, the more suicidal it is. But why does cancer evolve toward a form that will ultimately kill itself?

The principles of evolutionary biology offer insight. First, cancer behaves like a unicellular animal. Bacteria grown in a Petri dish will continue to grow until the food runs out. They make no effort to reduce growth in response to the obviously depleting food resources because each individual cell is concerned about only its own growth at that time. Grow until the food is gone, and then die. This is precisely the same growth pattern seen in cancer, which continues to grow until the host organism dies. At that point, the cancer must also die.

Second, cancer largely strikes older adults, long after the reproductive phase of life. Genes that increase cancer risk are still passed on to the next generation. For example, the *BRCA1* gene significantly increases the risk of both breast and ovarian cancer. The average age of diagnosis of breast cancer is 42.8 years, often after a woman has had children. Thus, despite the lethality of cancer, the *BRCA1* gene is passed on, and persists in the population.[10]

Cancer differs from other genetic diseases because it evolves. Sickle cell anemia, for example, is caused by a genetic mutation that is identical in all cases and stable over time and space. Cancer cells have mutations that change constantly—from person to person, and even within the same person over time. But if cancers are constantly and independently mutating, then how did they end up so similar, sharing all the same hallmarks? There are two possibilities: convergent evolution and atavism.

CONVERGENT EVOLUTION

Under similar environmental conditions, animals may independently evolve similar advantageous features in a phenomenon known as convergent evolution. For example, the flying squirrels of Australia and North America are genetically unrelated yet almost identical in appearance. These animals evolved similar features because they both faced the same natural selection pressures. Squirrels on both continents face ground predators, and winglike appendages that allow them to glide from tree to tree confer a huge advantage to survival. The two unrelated species converged upon the same solution: gliding.

Developing winglike flaps of skin requires significant genetic changes. If you were to ask, "Why did these squirrels develop gliding ability?" the answer could be genetic mutation. However, this is only the proximate cause. Ultimately, the *environment* selects the gene mutations that allow gliding. The genes of the two species are vastly different, but the mutations converged upon a similar phenotype, hence the name "convergent evolution." This forward evolution adds new abilities to the squirrels' existing frame—in this case, the ability to glide.

So, let's get back to cancer. Is this a case of convergent evolution? Every case of cancer in history must independently reinvent the wheel by adding new hallmarks in a stepwise manner. There is an infinite number of possible new mutations. Starting with a normal genome, the burgeoning cancer adds new mutations—to grow, to become immortal, to move, and to use the Warburg effect—step by step, until it becomes a full-fledged cancer.

But if each of the millions of cancers in history are evolving independently, how can they be so similar? It cannot be the environments, because they are utterly different. The lung is completely different from the breast, which is completely different from the prostate. How can every cancer look indistinguishable

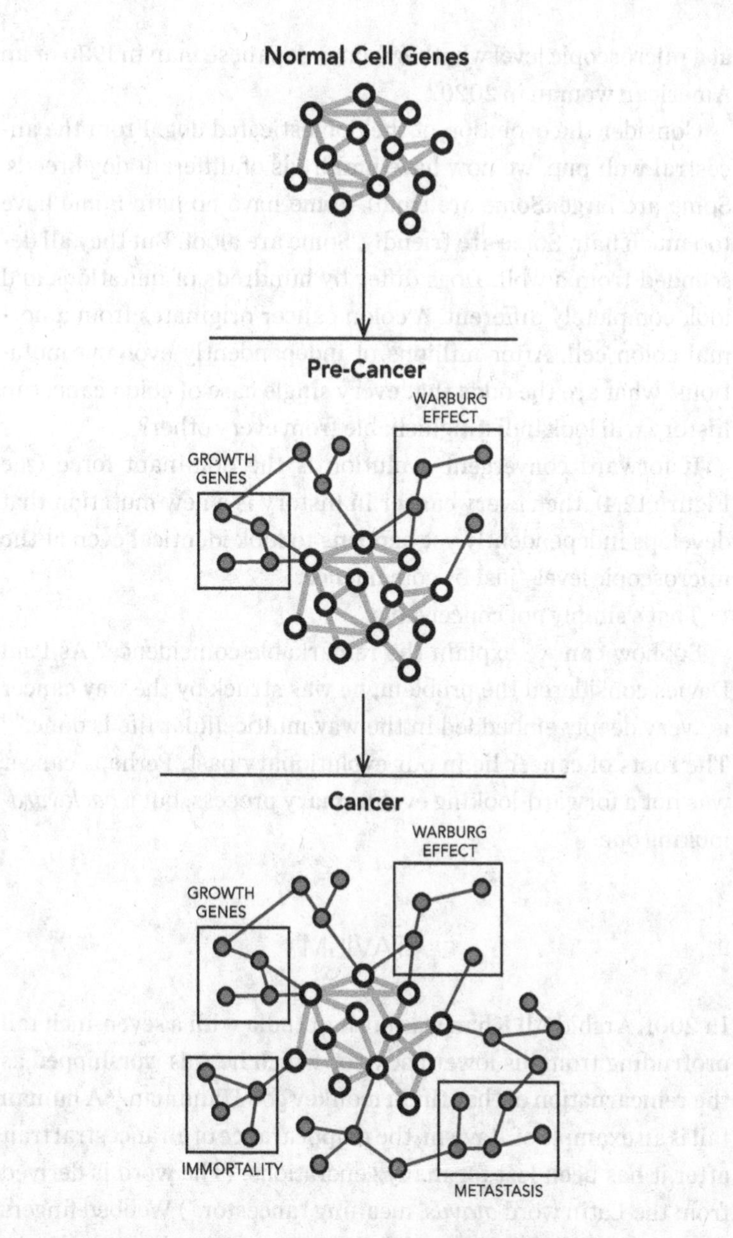

Figure 12.4: Forward evolution of cancer—sequential addition of mutations.

at a microscopic level whether from a Japanese man in 1920 or an American woman in 2020?

Consider the evolution of the domesticated dog. From the ancestral wolf pup, we now have hundreds of different dog breeds. Some are large. Some are small. Some have no hair. Some have too much hair. Some are friendly. Some are aloof. But they all descended from a wolf. Dogs differ by hundreds of mutations and look completely different. A colon cancer originates from a normal colon cell. After millions of independently evolving mutations, what are the odds that every single case of colon cancer in history will look indistinguishable from every other?

If forward convergent evolution is the dominant force (see Figure 12.4), then every cancer in history is a new mutation that develops independently yet happens to look identical even at the microscopic level—just by coincidence.

That's simply not conceivable.

So, how can we explain the remarkable coincidence? As Paul Davies considered the problem, he was struck by the way cancer is "very deeply embedded in the way multicellular life is done."[11] The roots of cancer lie in our evolutionary past. Perhaps cancer was not a forward-looking evolutionary process, but a *backward*-looking one.

ATAVISM

In 2001, Arshid Ali Khan was born in India with a seven-inch tail protruding from his lower back, for which he was worshipped as the reincarnation of the Hindu monkey god Hanuman.[12] A human tail is an example of atavism, the reappearance of an ancestral trait after it has been lost for many generations. (The word is derived from the Latin word *atavus*, meaning "ancestor.") Webbed fingers are another kind of atavism. While rare, atavisms occur regularly. But how did they develop? There are two general possibilities:

1. Hundreds of mutations come together to form a tail from scratch (de novo). This is a forward evolution, the addition of a new trait to an existing structure.
2. The biological plan for a tail already exists but is normally suppressed. Losing the suppressive mechanisms allows the tail to manifest. This is a backward evolution, the uncovering of an ancient but hidden trait.

The first possibility involves an unbelievable confluence of events. The muscles and connective tissue that form the tail must grow in a tubular shape. The overlying skin cells must grow to properly cover the tail. The blood vessels must grow to supply this anomalous tail. If this is a de novo mutation, then it does not necessarily need to look like a tail. It could look like an ear or a finger. It also does not necessarily need to develop where the normal tailbone sits. It could, for example, develop on top of your head or in your armpit.

The second possibility suggests that the human embryo already contains all the necessary genetic instructions for a tail, reflecting our evolution from primates. Humans evolved genes to suppress growth of the tail, but the original blueprints remain deeply buried. A rare breakdown of the tail-suppressing genes will allow the "tail-growing" genetic program to run because it has been there all along. When this genetic mutation occurs, every outward appearance looks identical to other people in history who have manifested a tail.

Imagine an art classroom where each child produced an identical picture of a flower—same size, same colors, same flower. Did every child just independently decide to paint the identical picture? Hardly. It's more likely that the picture was a paint-by-numbers kit that each child simply took it out of the box. In cancer's case, is it more likely that every cancer in history decided to evolve all the hallmarks independently, or that these hallmarks already existed and needed only to be uncovered?

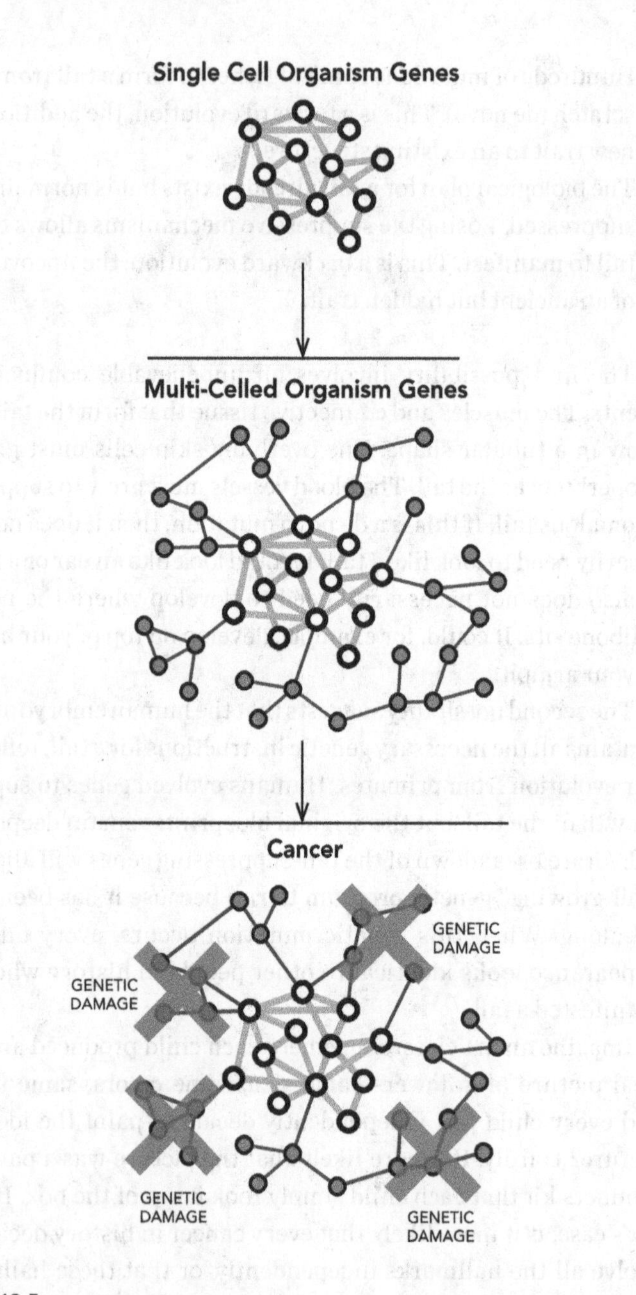
Single Cell Organism Genes
Multi-Celled Organism Genes
Cancer
GENETIC DAMAGE
GENETIC DAMAGE
GENETIC DAMAGE
GENETIC DAMAGE

Figure 12.5

The atavistic theory proposes that cancer is a reversion to an evolutionarily earlier format, the unicellular cell. *Cancer already exists buried deep into every cell of every multicellular animal.* This basic blueprint is already assembled, needing only to be uncovered. This atavism is essentially a backward, not a forward, evolution. It is a return to an earlier surviving version. This plausibly explains how every cancer in history develops independently but still looks the same.

Cancer is the unicellular ancestor of a normal cell. During the evolution to multicellularity, new control systems were added to the original program to ensure cooperation and coordination. Single-cell organisms grow, are immortal, move around, and use glycolysis. As multicellularity evolved, new genetic instructions were added to stop growth, make cells mortal, stop cells from moving around, and change energy generation to favor OxPhos.

But, critically, this old, unicellular programming *was not erased.* It still exists, albeit suppressed. New programs were simply built on top of the older ones. If the new suppressive programming fails, then the old programming can shine through. Atavisms are like pet tigers. You can train a tiger to tolerate humans and eat from its bowl. But if it becomes angry and forgets its training, the tiger reverts to being a wild animal.

The atavist theory suggests that unicellular organisms contain an original kernel of core genetic programming that allows growth, immortality, movement, and glycolysis. This kernel exists in multicell organisms as a remnant of the organism's evolutionary past as a single-cell organism. New genetic programming is layered over the old core to change the old competitive behavior to cooperative behavior. If these more recently added genetic controls are damaged, then the ancestral traits reassert themselves. That is how a normal cell completes the cancerous transformation (see Figure 12.5).

This theory makes the wild, but correct, prediction that cancer is a common, not a rare, event, as it is relatively simple to damage controls instead of building hundreds of coordinated new mutations in forward convergent evolution. It is the survival of the cell, not the multicellular organism, that is our ancient birthright. Indeed, the rate of cancer is far, far in excess of the known rate of mutation.

Cancer is guided along the same pathway backward toward unicellular life. Because this primitive programming exists in all cells, as cancers advance, they become more and more similar to one another. They dedifferentiate (become less different from one another). Indeed, *dedifferentiation* is the precise term used to characterize cancers' behavior.

Cancers all reach the same destination (unicellularity) by following a guided path (atavism) rather than a random walk (convergent evolution). Convergent evolution is about addition; atavism is about subtraction. These underlying pathways have evolved over millions or billions of years. Cancer already exists in all multicell organisms; it needs only to be revealed.

Why is cancer found in all multicellular animals? Why can every cell in the body become a cancer? Why are cancers so common? Why are cancers so similar to one another if they developed independently? The SMT had no answers, but atavistic evolution explains much of cancer's behavior. Still, what was the initial selection pressure that transformed a cooperative cell from a multicellular organism to a competitive single-cell organism? Now that we know *what* cancer is, we can ask a new question: what causes cancer?

13

CANCEROUS TRANSFORMATION

THE NEW EVOLUTIONARY paradigm of cancer was finding completely unexpected answers. Cancer is, improbably, a backward evolution, or atavism, toward the single-cell organism from our evolutionary past. Cells from multicellular organisms must suppress their unicellular urges. When these unicellular traits become exposed, the result is cancer. Is there any proof for this? Recent research has been finding more and more. According to this theory, cancer cells should express more ancient unicellular genes and fewer genes from the more recent multicellular period. This is exactly what studies are now finding. The number of cancer mutations peaks exactly at the intersection of unicellularity and multicellularity.[1]

A study in 2017 divided the 17,318 known human genes into sixteen different groups, called phylostrata, based on their evolutionary history (see Figure 13.1). The evolutionarily ancient genes in phylostrata one through three belonged to unicellular life. Phylostrata four through sixteen contained more recent genes.

Researchers then asked not which genes mutated, but *when*. From which evolutionary time period? Does the expression of older genes increase in cancer while the expression of more recently evolved genes decreases, as predicted by the evolutionary theory?

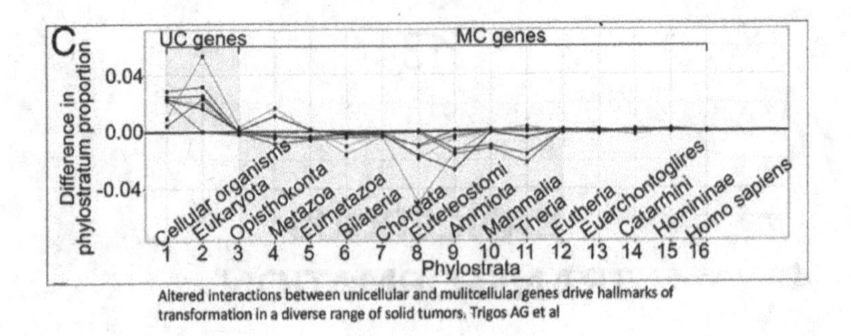

Figure 13.1

The answer is yes. Cancer preferentially expressed the ancient unicellular genes of phylostrata one through three. The genes representing the transition from unicellular to multicellular life, phylostrata four through eleven, were the most consistently and noticeably disrupted in cancer. These were the precise genes responsible for enhancing intracellular cooperation.[2] Cancer cells expressed more unicellular genes, which increase competition between cells. At the same time, cancer suppresses those genes that tried to control these unicellular urges and encourage cooperation. This study offers comprehensive evidence that cancer cells are trying to genetically move back toward unicellular life. The more aggressive the cancer, the more it expressed unicellular genes.

The Sanger Institute's COSMIC (Catalogue of Somatic Mutations in Cancer) database, the world's largest and most comprehensive collection of the various genetic mutations in cancer, was launched in 2004. The latest version from 2019, COSMIC Release v90, documented more than nine million different coding mutations.[3] A review of these mutations finds the same concentration of cancer genes at the onset of multicellularity.[4] Cancer preferentially mutates those genes that developed just after the onset of multicellularity, approximately five hundred million years ago. Cancer mutations are not random, but are targeted specifically at the junction of unicellular and multicellular life, exactly as the atavistic theory predicted.[5] Improbably, the wild prediction was 100 percent correct. This theory predicts that mutations that release the chains (tumor suppressor genes) are more important to cancer's genesis than those that provoke growth (oncogenes). In other words, it's easier to take your foot off the brakes than to build a new gas pedal. Once again, this is exactly what recent research has uncovered.

In clinical cancer samples, changes to tumor suppressor genes are 2.3 times more prevalent than oncogenes.[6] Cancer progressively demolishes the existing regulatory structures to reactivate its "genetic memory" of being a single-cell organism. The tumor suppressor gene *p53*, the most important in human cancer by far, is found in more than 50 percent of all cancers. The *BRCA1* gene, well known to increase risk of breast and ovarian cancer, is also a tumor suppressor gene.

Experimental tumor evolution studies found an astounding 12,911 genes that showed evidence of selection pressure, and more than 75 percent involved a reduction in gene expression.[7] Removing growth suppression is more important than the acceleration of growth genes. Logically, it is far easier to reduce the function of a gene than to increase its expression. If you randomly hit your car with a wrench, you're more likely to cause damage than to make it work better. Cancer is not about adding more functions; it is about subtracting control over existing functions. Cancer is less about a gene gaining new abilities than uncovering old ones.

Different cells, like a lung cell and a liver cell, have vastly different structures, functions, and environments. As cells evolve toward cancer, they lose specialized features and begin to look more and more alike. They become more *primitive* and *dedifferentiated*, two pathologic terms often describing cancers. Conceptually, cancers evolve toward the same unicellular destination: the stem cell.[8] This reversion toward a stem cell allows a lung cell to change sufficiently to live in the liver, because the primordial lung cell shares some common features with the liver. Cancer is a backward evolutionary process toward what can be easily considered a new species.

SPECIATION

Cancer cells are viewed by our immune system as a new invasive species. We are constantly exposed to microscopic foreign organisms, and the highly lethal cells of our immune system must carefully distinguish between "self" and "non-self" cells. We want to kill foreign invaders, not our own cells with friendly fire. Like viruses, bacteria, and fungi, cancer cells are recognized by natural killer cells as "non-self" and targeted for death.

Cancer may have evolved from a normal cell initially, but from the immune system's point of view, it has developed into a foreign species. In nature, speciation (the development of new species) is not a rare event. Dogs may have evolved from wolves, but they are not wolves. Breast cancer may have evolved from normal breast cells, but they are not normal breast cells. Cancer cells differ from the cells they are derived from in a number of ways, including in their being less specialized and being dedifferentiated.

Cancer can be considered an invasive species for several reasons: it redirects energy and resources toward itself, not the entire organism; it propagates and protects itself at the host's expense; and it adapts to survive in the hostile environment of the human body.[9] Cancer cells also evolve over time and space through an evolutionary path that is quite separate from that of the organism as a whole. A normal breast cell remains genetically the same, decade after decade. Breast cancer, however, contains multiple subpopulations of genetic variants that change over time.

These behaviors allow cancer to adapt to changing environments for survival. When we try to poison it through chemotherapy or burn it through radiation, the cancer evolves resistance in the same way that bacteria can develop antibiotic resistance. Where a cancer was originally derived from normal cells, it has diverged sufficiently to be considered alien. But what caused cancer in the first place?

WHAT CAUSES CANCER?

As I mentioned in the previous chapter, in a functional society, individuals must cooperate for mutual benefit. When a government fails, people will do whatever is necessary to survive and protect their families, resulting in anarchy. Desperate times call for desperate measures. A warlord emerges, ruling through sheer brutality. That warlord is cancer.

In cancer, the normal law and order of a multicell organism break down. The individual cells survive, but there are no rules for cooperation. To survive, cells must return to their old survival programming. The unicellular kernel, that most fundamental part of the cell that was born millions of years ago, is the ultimate survivor. The cell changes its behavior using ancient tools to ensure survival. Restraints on this old unicellular playbook are removed.

Behaviors typical of single-cell organisms return: growth, immortality, movement, and glycolysis. The cell has now completed its transformation to its evolutionary ancestor, the unicellular organism known as cancer.

Virtually any form of cellular or DNA damage may cause cancer, including chemicals, radiation, and viruses, but only under extremely specific conditions. The damage induced must be both:

- sublethal and
- chronic

To cause cancer, cell damage cannot be too much or too little. An overwhelming injury will simply kill all the cells, leaving no chance for cancer to develop. Dead people don't get cancer. If a city is completely destroyed by a nuclear bomb, nothing survives to compete for resources. Too little cellular damage, however, is simply repaired by the normal DNA-repair mechanisms. The immune system hunts down the occasional cancer cell, and all returns to normal. Carcinogenesis lies in between, like a man who has just missed his train—too late for one and too early for the next. In this gray zone between life and death, damaged cells try to survive, but the normal structures to ensure cooperation are no longer functional. Cancer is born in this struggle for existence.

Chronicity is the second key attribute of carcinogens. A single large dose of radiation is far less carcinogenic than chronic low-level radiation. The fallout from the atomic bombs dropped in Japan produced far less cancer than originally expected. A single large dose of cigarette smoke is far less carcinogenic than chronic cigarette smoking. The hepatitis A virus that causes a single episode of massive liver damage is far less prone to causing liver cancer than the chronic low-grade damage caused by hepatitis B or C. A single severe stomach infection is not carcinogenic, but chronic, low-grade stomach infection with *H. pylori* is.

Chronic sublethal damage activates cellular repair mechanisms, stimulating cell turnover and division. The only major difference between wound healing and cancer is that cellular growth eventually turns off when the wound is healed; not so with cancer. The striking similarity has led some researchers to dub cancers the "wounds that do not heal."[10] During wound healing, certain cellular attributes, like growth and immortality, are highly advantageous. Gene mutations in oncogenes and tumor suppressor genes such as *myc*, *PTEN*, and *src* that allow increased growth and replication (immortality) are highly beneficial in chronic wound healing and therefore accumulate slowly. This may form a premalignant lesion, such as a colonic polyp or the dysplasia seen in cervical cancer. Chronic, sublethal damage provides the time and continued selection pressure needed for cancerous transformation.

Carcinogenesis is an evolutionary process, and therefore takes time. A single acute injury does not exert the continuous selection pressure needed to cause cancer. It is the chronic chemical exposure, or the chronic radiation exposure, or the chronic infections, that cause cancer. Cancer also tends not to be an all-or-nothing proposition. When the selection pressure for growth and replication is removed, the risk of cancer recedes. For example, stopping smoking may reduce the excess risk of lung cancer by almost 75 percent after twenty years.[11]

Virtually any chronic sublethal injury may cause cancer. One of the clearest illustrations of this principle is found in the condition known as Barrett's esophagus. This is most commonly caused by gastroesophageal reflux disease (GERD), also known as reflux or, colloquially, heartburn. Normally, stomach acid stays in the stomach and does not back up into the esophagus. The lining of the stomach is designed to withstand the strong acids produced, but the lining cells of the esophagus are not. When stomach acid refluxes upward, the esophageal lining sustains damage, which causes the pain of heartburn. In response, the cell lining of the esophagus changes to one more closely resembling the stomach and intestines, in a process called metaplasia.

Barrett's esophagus is often considered a precursor to cancer, and has been increasing in recent decades. It converts to cancer of the esophagus at an annual rate of approximately 0.3 percent,[12] which is about five times higher than normal. The most significant risk factor for GERD and Barrett's esophagus is obesity.[13]

In this case, the cancer-causing agent is stomach acid, an entirely normal substance when in the right place. Stomach acid in the stomach is fine. Stomach acid in the esophagus is not fine, because the chronic sublethal cellular damage eventually leads to cancer.

All known carcinogens (tobacco smoke, asbestos, soot, radiation, *H. pylori*, and viruses) are chronic sublethal irritants. Ironically, some cancer treatments cause chronic irritation, and therefore cause cancer. Surgery is perhaps the oldest known cancer treatment. However, cancer may recur at the surgical site, even when all the surgical margins are clear. The trauma of the surgery itself induces chronic inflammation and wound healing, which facilitates the return of the cancer. In some rare cases, cancer may flourish at the site of an unrelated trauma. In one case, a patient became badly bruised after a fall. Two months later, he was diagnosed with lung cancer, which had metastasized to the site of the previous trauma.[14] This phenomenon is known as inflammatory oncotaxis.[15]

Radiation treatment burns cancer cells, and in a high enough dose, it may be curative. Go big or go home. But the treatment itself causes chronic sublethal cellular damage, and may therefore be carcinogenic. Secondary cancers develop in an estimated 13 percent of breast cancer patients, and the major risk is radiation treatment.[16]

Chemotherapy drugs are also well-known carcinogens. The chemotherapy drugs chlorambucil, cyclosporin, cyclophosphamide, melphalan, alkylating agents, and tamoxifen are all recognized as IARC group 1 carcinogens. Cyclophosphamide, an immune-suppressing medication used for autoimmune diseases like vasculitis[17] and rheumatoid arthritis,[18] is associated with up to a fourfold increase in certain types of cancer.

Today's standard anticancer treatments resemble ancient existential threats: radiation (pre-ozone layer), poison, and antimetabolites (nutritional challenges, periodic starvation). Unicellular cells are no strangers to these threats and have evolved effective responses to flourish under these precise conditions. This explains why the treatments from cancer paradigm 1.0 offered such limited benefit.

While carcinogenesis is a powerful force, so are the body's anticancer defenses. Multicellular animals have evolved a vast array of cancer-suppression mechanisms to maintain cellular law and order. This includes apoptosis (controlled cell death), DNA-repair mechanisms, DNA surveillance, epigenetic modifications, limited number of cell divisions (Hayflick limit), telomere shortening, tissue architecture, and immune surveillance. Mostly, these defenses against the dark arts are sufficient to keep us cancer free. But if environmental influences swing the advantage to the unicellular side, cancer may develop.

We've long treated cancer as some kind of random genetic mistake. A mistake that arises in all animals throughout history and evolves independently in millions of people a year? Cancer is hardly a mistake. Cancer is the ultimate survivor. When all else dies, cancer is there because it is the core of the cell that will survive at all costs. Cancer is not random, and it is not stupid. It has developed the tools it needs to survive.

This model fits the known facts of cancer better than any previous paradigm. Undoubtedly, this will not be the final word on cancer, and it should not be judged as such. Nor are all its suppositions proven facts. There will always be much to learn about cancer, but I believe this new paradigm is a huge and useful step forward, explaining many of cancer's mysteries.

EXPLAINING CANCER'S MYSTERIES

How Can Cancer Strike Every Part of the Body?

Most diseases target only one organ system. Hepatitis B attacks the liver, but not the foot. Alzheimer's disease attacks the brain, but not the heart. Cancer attacks every cell in the human body. Why? Because every cell in the body already contains the seed of cancer.

How Can Cancer Strike Virtually Every Multicellular Life-Form on Earth?

All animals and plants on Earth originated from single-cell organisms, so our genomes are preloaded with the "cancer subroutine," a deeply embedded, omnipresent set of genes. For single-cell organisms, of course, these are not directions to form cancer. Instead, they are simply instructions for how to successfully compete against other cells for dominance of their environment.

Multicellularity superimposed control procedures over these unicellular urges. The old "how to compete" playbook was not destroyed. Instead, new sections were added, changing it into a "how to cooperate" playbook. When these new pathways break down, the underlying unicellular (cancer) subroutine is unleashed; the old playbook is dusted off. Once this program is activated, it follows a predetermined script. The cancerous cell starts growing, forming a small mass of abnormal cells: the tumor.

Why Do All Cancers Look So Similar?

Cancers share the same "strange portfolio" of hallmarks, despite the widely varying cell type of origin and genetic differences among people. There is no a priori reason for these hallmarks to congregate together. Why should growth and immortality be selected along with the Warburg effect? Why wouldn't some cancers instead evolve the ability to photosynthesize energy from the sun? Why would cancer cause exuberant growth but not laser beams to shoot out of our eyes?

The cancer program is already predetermined. It is the reversion to the unicellular form of the cell, an atavism. All cancers share the same unicellular ancestor, with its ancient modus vivendi, a combination of attributes evolved over millions of years to maximize survival for the self.

Why Is Cancer So Common?

The lifetime risk of a clinical cancer of any site in humans is around one in three. In the United States, the lifetime risk of breast cancer in a woman is about one in nine. But the true incidence of cancer is many, many times that. Autopsy studies of people who die of non-cancerous causes indicate a strikingly high rate of undiscovered malignancies.[19] Cancer is not a rare disease; it is a ubiquitous one.

Each cancer must undergo hundreds or thousands of mutations to transform from a normal cell. The development of cancer, one mutation at a time, would take centuries if not millennia. The atavistic theory explains perfectly why cancer is so common: the origin of cancer already lies within every cell of our body. We don't need to build it up; we need only to uncover it.

A NEW UNDERSTANDING OF CANCER

In the next section of this book, we will look more closely at the tools we have at our disposal to prevent and fight cancer—shifting the focus from the seed to the soil. But before we do, I want to briefly summarize our journey to better understand the origins of cancer.

Cancer paradigm 1.0, as you'll recall, posited that cancer is a disease of excessive growth. This is certainly true, but it fails to explain *why* cancer cells grow so exuberantly. Cancer paradigm 2.0 suggested that cancer is a disease of genetic mutations that cause excessive growth. This, too, is certainly true, but again, it does not explain *why* these mutations occur. This led to cancer paradigm 3.0, the evolutionary theory, which postulates that genes mutate as a survival response against chronic sublethal injury. The drive toward a unicellular state, the most basic unit of survival, is the force behind the mutations. Charles Darwin's theory, as applied to cancer, is best summed up by the paraphrase "It is not the strongest species that survive, nor the most intelligent, but the ones most responsive to change."[20] Cancer is the ultimate cellular changeling fighting for its own survival. This powerful enemy has been shaped by the most potent known biological force: evolution.

Cancer behaves like an invasive species. The relentless growth and eventual metastasis of cancer reflects an organism seeking out a new environment to thrive. Where cancer paradigm 2.0 considers cancer as a "dumb" disease of genetic mutations, cancer paradigm 3.0 considers it to be a "smart" disease, an invasive protozoan doing everything in its power to survive.

Cancer is an ever-present danger because its seed lies in every cell in every multicellular organism. Like aging, it can never really, truly be eradicated, but tilting the odds is certainly possible.

Cancer moves through three phases: transformation, progression, and metastasis. What we've described so far is only the first step: cancerous transformation. Evolution has worked at cancer like a sculptor works at stone, designing, chiseling, rounding, edging, and perfecting its survival genes. Gradually, the finished piece emerges: a lethal work of art. But what are the environmental conditions that allow cancer to flourish?

PART IV

PROGRESSION

(Cancer Paradigm 3.0)

14

NUTRITION AND CANCER

IN 1981, THE U.S. Congress Office of Technology Assessment asked Sir Richard Doll, the preeminent cancer epidemiologist of his era, and Sir Richard Peto, an esteemed professor of medical statistics and epidemiology at Oxford University, to estimate the known root causes of cancer. Their 117-page landmark document[1] was updated in 2015, and overall, researchers agreed that the original estimates would "generally [hold] true for 35 years."[2]

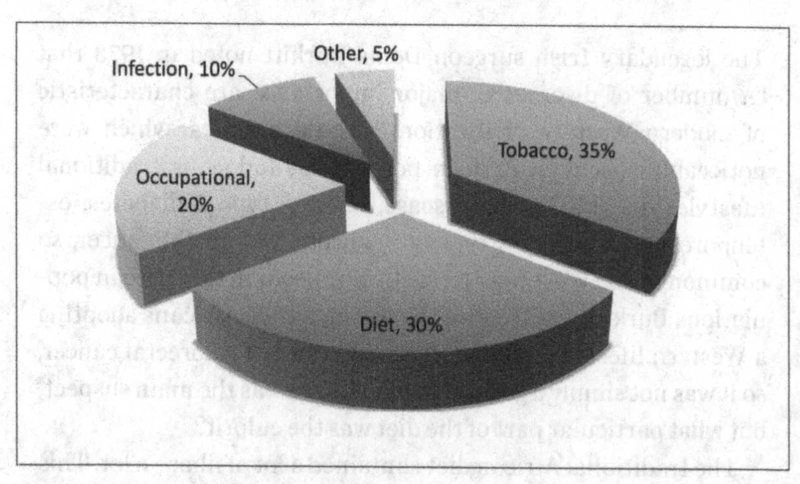

W. J. Blot and R. E. Tarone, "Doll and Peto's Quantitative Estimates of Cancer Risks: Holding Generally True for 35 Years," *Journal of the National Cancer Institute* 107, no. 4 (2015): djv044.

Figure 14.1

Tobacco was, and still is, the most important contributor to cancer. Smoking in the United States peaked in the 1960s, although about 20 percent of the adult population still smokes, accounting for approximately one third of the known risk of cancer (see Figure 14.1).

Dietary factors, including obesity and inactivity, are a very close second, at 30 percent attributable risk of cancer, although they may plausibly account for up to 60 percent of risk. It was clear that the link between diet and cancer was a singularly important one, but the question remained: what specific dietary factor is responsible for cancer? Pinning down the precise relationship is incredibly difficult. Was it some kind of vitamin deficiency? Were we lacking some crucial protective factor? Was the problem too much of something in the diet, or not enough?

DIETARY FIBER

The legendary Irish surgeon Denis Burkitt noted in 1973 that "A number of diseases of major importance are characteristic of modern Western civilization."[3] These diseases, which were noticeably absent in African populations following traditional lifestyles, included heart disease, obesity, type 2 diabetes, osteoporosis, and certain types of cancers. Colorectal cancer, so common in the West, was virtually unknown in the African populations Burkitt was treating. However, native Africans adopting a Western lifestyle suffered increased rates of colorectal cancer, so it was not simply a genetic problem. Diet was the main suspect, but what particular part of the diet was the culprit?

The traditional African diet contained a lot of fiber—a lot. This bulked up the stool, leading to frequent and large-volume bowel movements. The Western diet contained little fiber because of the predominance of refined grains, which lost most of their natural fiber during processing. The result? Fewer, smaller bowel movements.

Fiber is the indigestible part of plant foods; it is either water soluble or insoluble. In both cases, it is not absorbed by the body, and adds bulk to the stool. Burkitt, putting two and two together, hypothesized that the high dietary fiber of the traditional African diet prevented colon cancers. Perhaps the regular movement of the stool cleared the intestinal system, preventing decaying and putrefying of foods inside the colon, which might be carcinogenic. The high stool volume meant frequent "cleansing" bowel movements. Eating more fiber was enthusiastically championed as an easy way to improve health and reduce cancer.

It was a pretty good hypothesis. However, the early studies on precancerous lesions, known as adenoma or polyps, were not encouraging. In the mid-1990s, two large trials, the Toronto Polyp Prevention Trial[4] and the Australian Polyp Prevention Project,[5] failed to detect any health benefits to eating more fiber.

By 1999, the Nurses' Health Study, a sixteen-year study involving more than sixteen thousand women, found that a high-fiber diet did not reduce the risk of precancerous adenomas. Yes, there were all those glorious bowel movements, but no, it did not lower the risk of cancer.[6] There was more bad news to come.[7] In a randomized controlled trial, the gold standard of experimental medicine, 1,303 patients were assigned to eat either their usual diet or a low-fat, high-fiber diet emphasizing fruits and vegetables. The high-fiber group ate a whopping 75 percent more fiber and 10 percent less fat than in a standard diet. Unfortunately, from a cancer-prevention standpoint, this dietary intervention was essentially useless. Yes, fiber makes your bowel movements better, but it does not prevent colorectal cancer.

Eating lots of dietary fiber was not the protective factor against colorectal cancer. It must have been one or more of the many other dietary and lifestyle factors that differed between the Africans and the Europeans. The Inuit of the Far North, following their traditional lifestyle, ate little or no fiber, as little plant material grew in the region. But they, too, had little or no colorectal cancer. Cancer was not simply a disease of too little fiber, and thus, eating more fiber didn't translate into less cancer. Rats!

DIETARY FAT

The next suspect was dietary fat, particularly saturated fats. There was no real reason to suspect that dietary fat should cause cancer. After all, humans had been eating fats, including saturated fats such as animal fats (e.g., meat, dairy) and plant fats (e.g., coconut oil, olive oil) for millennia. Traditional societies often ate large amounts of fat. The Inuit ate whale and seal blubber. Peoples of the South Pacific islands ate large amounts of coconut, which is high in saturated fat. Neither high-fat-eating population suffered particularly from cancers, heart disease, or obesity. There was simply no suggestion that dietary fat was carcinogenic in any manner. Why did we even consider this a reasonable possibility?

From the late 1960s to the 1990s, we were gripped in a hysterical fat phobia. After World War II, seemingly healthy middle-aged men were suffering heart attacks at an alarming and increasing rate. But nobody knew why. When President Eisenhower suffered a heart attack, it suddenly became the most important medical topic of its day. Obesity, type 2 diabetes, and lack of exercise were not major health issues yet. What was the culprit?

The most significant lifestyle changes from 1900 to 1950 were not dietary, but rather, the widespread adoption of cigarette smoking, a trend that accelerated after World War II. The connection between smoking and disease was obscured for decades as tobacco companies strenuously denied that the habit caused heart disease, lung disease, or cancer. Indeed, in the 1960s, doctors were happily puffing away with the rest of their generation.

Dr. Ancel Keys, a prominent nutrition researcher, pointed to dietary fat as the villain causing heart disease. This never made any sense. Americans, living in the land of plenty, had always eaten more animal fats than virtually anybody else in the world. The huge farmlands of the Midwest provided feed for the vast cattle ranches of Texas. Americans had always consumed a lot of beef and milk. Even at a superficial glance, it is hard to see how anybody could have concluded that eating more fat caused more heart disease. Consumption of fat was not rising, yet incidence of heart disease was increasing at an alarming rate.

But every story needs a villain, and dietary fat became nutritional public enemy number one. The American Heart Association (AHA) wrote the world's first official recommendations in 1961, suggesting that Americans reduce their intake of total fat, saturated fat, and cholesterol. Following that advice, people started to drink low-fat dairy products and switch from eating eggs and meat to consuming low-fat foods such as white bread and pasta.

But the anti-fat crusade didn't stop at heart disease. The phantom menace of dietary fat was blamed for almost everything bad. It caused obesity. It caused high cholesterol. It caused heart disease. It probably caused bad breath, hair loss, and paper cuts, too. There was no actual proof that dietary fat, which humans had been eating since . . . well, since we became human, caused harm. But it didn't really matter, because the entire scientific world had hopped on the fat-is-bad bandwagon. Because they thought dietary fat caused heart disease, they logically concluded that it probably caused cancer, too. Who needs proof if you have dogma?

Still, nobody had any idea *how* dietary fat caused cancer. Even anecdotally, there were few observations suggesting that people who ate a lot of fat got a lot of cancer. Not only have the Inuit and South Pacific Islanders, two populations with low cancer rates, been eating plenty of fat for centuries, but the vegetarians of India were eating very low-fat diets of mostly grain and weren't protected from cancer. It didn't matter. Blame-dietary-fat-for-everything-bad was the name of the game. So, play on!

In 1991, the Women's Health Initiative, an enormous randomized controlled trial, tested the theory that dietary fat caused not only weight gain and heart attacks but also breast cancer. Close to fifty thousand women took part, with one group of women instructed to follow their usual diet and the other instructed to reduce their dietary fat to 20 percent of calories and increase their intake of grains, vegetables, and fruits.

Over the next eight years, these women faithfully reduced their dietary fat and their overall caloric intake. Did this strict diet reduce the rates of heart disease, obesity, and cancer? Not even close. Published in 2007, the study found *no benefit* for heart disease.[8] The women's weight was unchanged. And their rates of cancer—well, they were no better, either. Specifically, there was no benefit for breast cancer,[9] and there was no benefit for colon cancer.[10] It was a stunning defeat. Did dietary fat play a tiny role in causing cancer? The answer was irrelevant. The effect was so small as to be undetectable, even by the biggest nutritional trial in history.

Lowering dietary fat resulted in no measurable health benefits, directly contradicting the prevailing beliefs of the time. Eating more dietary fat did not cause cancer. Eating less fat did not protect against cancer. In terms of causation of cancer, dietary fat was a dud. So, what was next?

VITAMINS

Could it be that cancer is caused by vitamin deficiency, and if so, could supplements reduce the risk of cancer? Vitamins are big business. People love taking supplements. It is a beautiful dream that simply taking some vitamins can reduce our risk of cancer. So, we put them to the test, but the results were not good.

Beta-carotene

First up was beta-carotene, a precursor to vitamin A that gives carrots their orange color and serves as a potent antioxidant in the body. A randomized controlled trial in 1994 asked whether beta-carotene supplementation could lower heart disease and/or cancer.[11] Hopes were high, but unfortunately, the answer was no. Cancer was not simply a beta-carotene-deficiency disease in the way that scurvy is a vitamin C–deficiency disease. Bad news for vitamin lovers.

But there was worse to come: taking beta-carotene supplements actually *increased* the rates of both cancer and overall death. At first, this effect was considered a fluke, but a similar 1996 study found the same cancer-causing effect.[12] How could taking seemingly benevolent vitamins worsen cancer risk? It would be a few more years before this mystery was solved: Cancer behaves like an invasive species. Vitamin supplementation benefits these fast-growing cells more than slower-growing normal cells.

Cancer cells grow and grow and grow and never stop. But not even cancer can grow without nutrients. Erecting a brick wall without any bricks is impossible, even for the best builders in the world. The rapidly growing cancer cells require a constant supply of nutrients. Vitamins don't cause a cell to turn into a cancer, but if a cancer is already present, they can certainly help it grow. Feeding cancer lots of vitamins is like sprinkling fertilizer onto an empty field in the hope of getting a nice, lush lawn. You want the grass to grow, but the weeds, being the fastest-growing plants in the field, also take up the nutrients and grow like . . . well, weeds. When vital nutrients like beta-carotene are available in large doses, cancer cells are highly active and grow like weeds. In cancer medicine, you don't want *more* growth; you want less.

Folic Acid (Vitamin B_9)

Next up was folate, the water-soluble B vitamin found in leafy greens, legumes, and cereals. It is so important for proper cell growth that the United States mandates that folate supplements be added to fortify flour. Folate supplementation was one of the more spectacular success stories of the modern era. Even in well-nourished Western societies, routine supplementation of pregnant women with folate (folic acid) significantly reduced the incidence of neural tube birth defects. A flurry of observational studies in the 1980s and '90s suggested that a low-folate diet increased the risk of heart disease and colorectal cancer. A huge wave of enthusiasm for B vitamin supplementation in the early 2000s led to large studies to see if we could reduce those diseases.

The 2006 HOPE2 randomized trial unfortunately found that both folate and vitamin B_{12} supplementation failed to reduce heart disease.[13] But what about cancer? The study saw a worrying trend toward increased risk of both colon (36 percent increased risk) and prostate cancer (21 percent increased risk) with supplementation. By itself, this was a big red flag, but more bad news was yet to come. The Aspirin/Folate Prevention of Large Bowel Polyps clinical trial found that six years of folate supplementation *increased* the risk of advanced cancer by 67 percent.[14] Another study found that breast cancer patients using vitamin B_{12} supplements had a higher risk of recurrence and death.[15]

Two large trials, the Norwegian Vitamin (NORVIT) trial[16] and the Western Norway B Vitamin Intervention Trial (WENBIT),[17] confirmed that high-dose B vitamin supplements did not reduce heart disease. Cancer? There was a significant effect, but not a good one. Folate supplementation *increased* the risk of cancer by 21 percent and cancer death by 38 percent.[18] Not good. We weren't preventing cancer; we were giving it to patients.

Winston Churchill once reminded us that "Those who fail to learn from history are condemned to repeat it." If we had only remembered a bit of the medical history, this sorry chapter in medicine might have been avoided. In 1947, the father of modern chemotherapy, Dr. Sidney Farber, tested folate supplementation in ninety human patients with incurable cancer.[19] Some cases, particularly the leukemia patients, responded with a marked *acceleration* of cancer growth. Patients got worse, not better.

The mark of a truly great scientist, though, is the ability to change one's mind when the facts change. Realizing that folic acid worsened cancer, Farber switched to administering aminopterin, a folate blocker, reasoning correctly that if folate was making patients worse, then blocking it might make them better. This seminal discovery launched the modern era of chemotherapy. Leukemic patients showed almost miraculous, although temporary, improvement.

Modern chemotherapy was based on the single observation made in the 1940s that giving folate *worsened* cancer. Yet, in the early 2000s, millions of cancer research dollars were spent to prove a fact well-known for decades. With hindsight, it's not difficult to figure out why high-dose vitamin supplementation worsens cancer risk. Cancer cells grow like crazy. High-dose vitamins promote cell growth. It is really just that simple.

Vitamin C

Deficiency of vitamin C causes the disease of scurvy, which was a constant danger for sailors in the years AD 1500 to 1800. The long voyages without adequate vitamin C resulted in easy bruising, swollen extremities, and inflamed gums. This disease was eventually cured by providing sailors with oranges and lemons to eat during the long sea journeys. Vitamin C could cure scurvy, but could it cure cancer?

In the 1970s, Linus Pauling—the only person to have won two unshared Nobel Prizes (in Chemistry, as well as the Peace Prize)—became a vocal proponent of vitamin C supplementation, believing it could cure the common cold and cancer, too.[20] Unfortunately, more recent studies have failed to confirm the anticancer effect of vitamin C in humans. A 2015 review of all available studies concluded that "there is no evidence to support the use of vitamin C supplements for prevention of cancer."[21] Vitamin C supplementation is important if you are a pirate of the Caribbean, but it cannot prevent or treat any form of cancer.

Vitamin D and Omega-3 oils

It was 1937 when scientists first speculated that sunlight exposure might lower the risk of cancer.[22] By 1941, studies were finding that living at higher latitudes (with less sunlight) was associated with a higher risk of cancer death.[23] This link could be explained by a protective effect of vitamin D, which is produced in the skin when it is exposed to ultraviolet B (UVB) rays from the sun.[24]

Sunlight exposure is the only significant source of vitamin D for most people, as few foods naturally contain significant amounts of vitamin D. Increasing UVB exposure might increase the risk of melanoma, so a promising alternative strategy was vitamin D supplementation, which gained significant popularity by the mid-2000s.

Animal and human studies[25] hinted at the vast potential for vitamin D as an anticancer agent.[26] Vitamin D could reduce cancer cell proliferation, increase apoptosis (a key anticancer defense), reduce new blood vessel formation, and decrease a tumor's invasiveness and its propensity to metastasize.[27] The Third National Health and Nutrition Examination Survey (NHANES),[28] a large American study involving more than thirteen thousand adults, found that low vitamin D levels were associated with a stunning 26 percent increase in total mortality, primarily from heart disease and cancer, the two biggest killers of Americans.

To find some definitive answers, the NIH randomized more than twenty-five thousand participants in a massive trial on vitamin D supplements and another popular supplement at the time, marine-derived long-chain n-3 (also known as omega-3 fatty acids) over 5.3 years. Unfortunately, the Vitamin D and Omega-3 Trial (VI-

TAL) found no evidence that either supplement prevented cancer in any way. Not for breast cancer. Not for prostate cancer. Not for colon cancer. Similarly, there were no benefits for the prevention of heart disease.[29] This lack of benefit was confirmed in 2018 by the Vitamin D Assessment Study (ViDa).[30] There were no harmful effects detected, but neither were there any benefits.

Vitamin E

Vitamin E is a group of fat-soluble antioxidant vitamins that became very popular in the 1990s, gaining a reputation for potentially reducing heart disease and cancer.[31] Unfortunately, large-scale randomized trials since then have found that vitamin E supplementation did not reduce the risk of colon,[32] lung,[33] prostate, or total cancers.[34]

Once again, there was a whiff of danger. The large-scale Selenium and Vitamin E Cancer Prevention Trial (SELECT) in 2009[35] found no benefit to reducing prostate cancer. But longer follow-up led researchers to conclude that "Dietary supplementation with vitamin E significantly increased the risk of prostate cancer among healthy men."[36] This was just ugly. Vitamin E supplements were causing the very cancers they were supposed to be preventing.

HOW *NOT* TO CURE CANCER

Cancer is simply not a disease of nutrient deficiency, and therefore taking supplements is not likely to make much difference. We tested vitamin A (beta-carotene), which failed to reduce cancer. We tested B vitamins, which failed. Then we tested vitamins C, D, and E . . . which all failed. We were running out of letters to test! So, here's what we are left with:

- Diet plays a large role in cancer.
- Cancer is not caused by lack of dietary fiber.

- Cancer is not caused by too much dietary fat.
- Cancer is not caused by a vitamin deficiency.

While it may sound trivial, these four bits of knowledge cost, literally, hundreds of millions of dollars of research money and many decades of work. This left unanswered the most important question of all: what component of the diet *is* responsible for so much cancer?

Beginning in the late 1970s, one nutritional measure began to eclipse all others in importance: obesity. The obesity epidemic began as an American phenomenon, but it has since gone global. Cancer is increasingly known as an obesity-related disease, accounting for 20 to 30 percent of the risk of common cancers.

OBESITY

Obesity is estimated to have caused 4.5 million deaths in 2013 worldwide, due primarily to the increased risks of both heart disease and cancer.[37] It is often clinically assessed by a measure known as the body mass index (BMI), which is one's weight in kilograms divided by their height in meters squared, or $BMI = kg/m^2$.

It's important to keep in mind that this simple calculation does not consider many factors, such as body composition (including muscularity and bone density), and as such, it is a flawed metric at the individual level. But over large populations, it is a reasonably useful measure. A generally accepted classification is as follows:

BMI < 18.5	Underweight
BMI 18.6–24.9	Normal weight
BMI 25–29.9	Overweight
BMI 30–39.9	Obese
BMI > 40	Morbidly obese

The obesity epidemic in the United States accelerated in the late 1970s, followed about ten years later with rising rates of type 2 diabetes. It was not until the 2000s that most researchers realized that obesity influences cancer, too. Because cancer often requires decades to develop, the worsening obesity crisis was just then making its presence felt.

The Cancer Prevention Study II, a large prospective cohort study, began in 1982. This massive scientific undertaking required 77,000 volunteers simply to *enroll* all the participants, which numbered over 1 million. The participants (average age: fifty-seven) were healthy and free of any detectable cancer at the beginning of the study. Every two years, they were tracked to see who had died and why.[38] In 2003, the data reached a then-novel and stomach-churning conclusion: obesity, already a well-known risk factor for diabetes, heart disease, and stroke, also significantly raised the risk of cancer.

The risk of cancer began to rise at a BMI greater than 30 (obese) and accelerated in those with a BMI over 40, who suffered 52 to 62 percent more overall cancer death (see Figure 14.2). The risk differed by cancer site. The risk of liver cancer is increased by 452 percent, and the highly lethal pancreatic cancer risk is increased by 261 percent.

Increased Cancer Risk with Obesity

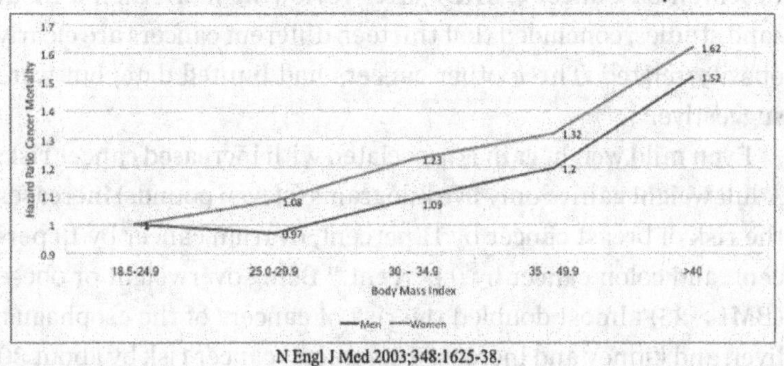

N Engl J Med 2003;348:1625-38.

Figure 14.2

This is devastating news, but the truth is almost certainly worse for two reasons. People found to have lung cancer from smoking tend not to be obese because of smoking's slimming factor. Because lung cancer is the most significant cause of cancer deaths, this means that the 52 to 62 percent increased risk of cancer with obesity is almost certainly an underestimate. Looking at nonsmokers only, the increased risk of cancer for a BMI over 40 is a disastrous 88 percent.

Second, advanced cancer patients tend to lose weight, a phenomenon known as cancer cachexia. This will similarly obscure the true link between obesity and cancer and, once again, lead to an underestimation of obesity's true risk.

Over the ensuing years, the news on obesity and cancer has only gotten worse. In 2017, the CDC released a report highlighting the "Trends in Incidence of Cancers Associated with Overweight and Obesity: United States, 2005–2014."[39] Some of the most common cancers, including breast and colorectal, are related to obesity and excess body fatness (see Figure 14.3). Together, these accounted for an astounding 40 percent of all cancers. The specific cancers most tightly linked to obesity are liver, endometrial, esophageal, and kidney cancers, with two to four times the risk. Breast and colorectal cancer are more moderately associated, at one and a half to two times the risk. In 2016, the International Agency for Research on Cancer (IARC), after reviewing more than a thousand studies, concluded that thirteen different cancers are clearly obesity-related. Three other cancers had limited data, but were suggestive.[40]

Even mild weight gain is associated with increased cancer risk. Adult weight gain of only five kilograms (eleven pounds) increases the risk of breast cancer by 11 percent, ovarian cancer by 13 percent, and colon cancer by 9 percent.[41] Being overweight or obese (BMI > 25) almost doubled the risk of cancers of the esophagus, liver, and kidney and increased colorectal cancer risk by about 30 percent.

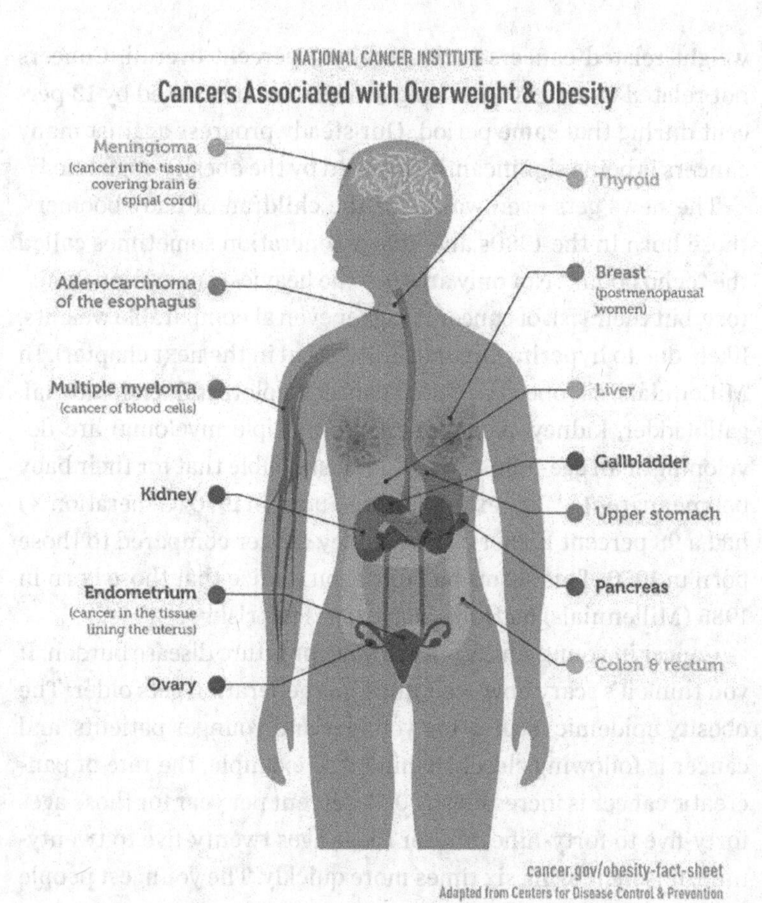

Figure 14.3

From 2005 to 2014, the incidence of all weight-related cancers actually dropped slightly, but a closer look at the numbers tells a vastly different story. This improvement was limited to only one single type of cancer: colorectal. Increased screening with colonoscopy detected and removed precancerous adenomas before they could turn into full-blown cancer, dropping colorectal cancer rates by 23 percent. But excluding colon cancer, the other weight-related cancers increased by 7 percent overall. Cancers not related to weight (like lung cancer) had decreased by 13 percent during that same period. Our steady progress against many cancers is being significantly impeded by the obesity epidemic.[42]

The news gets even worse for the children of baby boomers, those born in the 1980s and '90s, a generation sometimes called the "echo boom." Not only are they the heaviest generation in history, but their risk of cancer is higher even at comparable weights, likely due to hyperinsulinemia (discussed in the next chapter). In Millennials, six obesity-related cancers (colorectal, endometrial, gallbladder, kidney, pancreatic, and multiple myeloma) are developing at an age-adjusted rate almost double that for their baby boomer parents.[43] For example, those born in 1970 (Generation X) had a 98 percent higher risk of kidney cancer compared to those born in 1950. This seems bad until you realize that those born in 1985 (Millennials) had almost *five times* the risk!

Cancer in young adults is a warning for future disease burden. If you think it's scary now, wait until this generation gets older. The obesity epidemic is affecting younger and younger patients, and cancer is following closely behind. For example, the rate of pancreatic cancer is increasing by 0.77 percent per year for those ages forty-five to forty-nine, but for those ages twenty-five to twenty-nine, it is increasing six times more quickly. The youngest people are facing the steepest rises in cancer rates. By contrast, most cancers *not* related to obesity are decreasing, especially those related to viruses, smoking, and HIV.

If weight gain increases cancer risk, does weight loss lower it? The first animal studies that suggested this possibility were published over a century ago, by 1914 Nobel laureate Peyton Rous. In mice, severely restricting food availability cut their risk of cancer by *half*.[44] In the 1940s, Dr. Albert Tannenbaum, former president of the American Association for Cancer Research, discovered that, astoundingly, carbohydrate restriction alone in mice provided greater protection against cancer than overall calorie restriction.[45] He concluded that "tumor formation is dependent on the *composition* of the diet, as well as the *degree* of caloric restriction," a remarkably prescient observation.[46] In the Nurses' Health Study, women who lost ten kilograms or more after menopause and kept the weight off lowered their risk of breast cancer by an astounding 57 percent.

Obesity clearly increases the risk of cancer. Obesity also clearly increases the risk of type 2 diabetes. What is the link? The master hormone of metabolism: insulin.

15

HYPERINSULINEMIA

TYPE 2 DIABETES AND CANCER

Among the most prominent of Denis Burkitt's "diseases of civilization" are obesity, type 2 diabetes, and cancer, whose threads are so closely interwoven that they are almost indistinguishable from one another. Burkitt considered them different manifestations of the same underlying problem. But what was this unifying thread?

Populations that transition from traditional lifestyles and diets to Western ones suffer increased obesity, type 2 diabetes, and cancer. This connection, so obvious to Burkitt in Africa, has also been observed worldwide in the native populations of North America, the Inuit people of the Far North, the aboriginals of Australia, and the South Pacific Islanders. The largest group of indigenous people of the Arctic regions are the Inuit. Early reports from 1936 found no evidence of cancer in Inuit peoples.[1] None. The scientific literature of the time considered the Inuit virtually immune to both type 2 diabetes and cancer.

One 1952 expedition noted that "It is commonly stated that cancer does not occur in the Eskimos, and to our knowledge no case has so far been reported."[2] Living conditions in the Far North changed drastically after World War II. Improved housing in larger urban communities increased overall life expectancy. The traditional diet, based on fish and sea mammals, now relied upon imported foods—largely, refined carbohydrates and sugar. Soon came the realization that the Inuit were not, in fact, immune to cancer, obesity, or type 2 diabetes.

The story of type 2 diabetes in the Inuit is eerily similar to that of cancer. At the turn of the twentieth century, Arctic medical expeditions searched in vain for the secret to the Inuit's "immunity" to type 2 diabetes. One researcher wrote in 1967 that "Clinical experience has long suggested that diabetes mellitus is rare in Alaskan Eskimos."[3] The story began to change in the 1970s, and not for the better. By 1988,[4] surveys showed that "Diabetes mellitus is no longer a rare condition among Alaska Natives."[5] From 1990 to 1998, diabetes increased by 71 percent among native Alaskan children and young adults.[6]

Not coincidentally, the rates of obesity, type 2 diabetes, and cancer in native populations throughout the globe have soared in tandem. From nonexistent diseases in people following a traditional lifestyle to an epidemic, all within the space of a single generation. Disaster. Because of its clear correlation to changes in lifestyle, cancer had long been considered a "Western" disease, not a genetic one.

Dr. D. G. Maynard first noted the increased risk of cancer with diabetes in 1909.[7] Both diseases were relatively obscure (oh, how things change) and rising in incidence (oh, how things stay the same). Since then, this correlation has been confirmed many times over in the scientific literature. Like the closely associated dietary disease of obesity, type 2 diabetes consistently increases the risk of cancer.

In the United States, the 2012 Cancer Prevention Study II estimated that diabetes increases the risk of cancer death by 7 to 11 percent.[8] A 2011 European analysis[9] and a 2017 Asian study[10] both estimated that diabetics have about a 25 percent greater risk of cancer death. Each increase of blood glucose by 1 mmol/L was associated with an increased risk of fatal cancer: 5 percent for men and 11 percent for women.[11]

As in the case of obesity, diabetes increases the risk of certain types of cancer more than others. A consensus conference from the American Diabetes Association estimates that diabetes doubles the risk of liver, endometrial, and pancreatic cancer and increases the risk of breast and colorectal cancer by approximately 20 to 50 percent.[12] This may largely explain why the age-adjusted rate of pancreatic cancer rose by 24 percent in the United States between 2000 and 2014.[13]

The close association of obesity, type 2 diabetes, and cancer clearly establishes cancer as a disease of metabolism, rather than one purely of genetic mutations. It is a seed and soil problem. When the seed of cancer is planted, the metabolic milieu of the body allows it to flourish. But what links these three related diseases? The hormone insulin.

Obesity is primarily a disease of hyperinsulinemia. The word *hyperinsulinemia* is derived from the prefix *hyper*, meaning "excess," and the suffix *-emia*, meaning "in the blood." Thus, *hyperinsulinemia* is translated literally as "too much insulin in the blood." Obesity is commonly mistaken for a disease of excessive calorie intake, but it is in fact mainly a hormonal disorder of excessive insulin. (The details are beyond the scope of this book, but if you'd like to learn more about them, I offer an in-depth discussion of this phenomenon in my book *The Obesity Code*.) Type 2 diabetes is similarly a disease of hyperinsulinemia, although in this context it is often termed "insulin resistance." Like Superman and Clark Kent, hyperinsulinemia and insulin resistance are one and the same.

So, let's recap:

- Obesity is a disease of hyperinsulinemia.
- Type 2 diabetes is a disease of hyperinsulinemia.
- Is cancer also a disease of hyperinsulinemia?

Insulin is a natural hormone released into the blood by the pancreas in response to food—specifically, dietary protein or carbohydrates, but not dietary fat. Generally, insulin functions as a nutrient sensor because it signals the availability of carbohydrates and proteins to the rest of the body, which may then either use or store this food energy. Is hyperinsulinemia carcinogenic? Thirty years ago, the very thought would have been widely ridiculed. Today, it is one of the hottest areas of cancer research.

INSULIN AND CANCER

The cancer-provoking potential of insulin was noted as early as 1964.[14] In lab cultures, normal breast cells incubated with insulin proliferated with such enthusiasm that they resembled cancers. Growing breast cancer cells in the laboratory also requires insulin. Lots of insulin. This is an interesting observation because normal breast cells don't really need insulin. Yet breast *cancer* cells can't live without it.[15] If you remove the insulin from breast cancer cell culture, the cells quickly shrivel up and die. This holds true for other cancers like colorectal, pancreatic, lung, and kidney, too. In mice, injecting insulin induces the growth of breast and colon cancer.[16]

This was a puzzling anomaly. The main tissues normally involved with glucose metabolism, the liver, fat cells, and skeletal muscles, naturally have the highest numbers of insulin receptors. Normal breast tissue? Not so much. So, why does breast cancer thrive on insulin? Breast cancer cells express six times the levels of insulin receptors compared to normal breast cells.[17]

High insulin levels can be measured with a blood test called the C-peptide, which is a protein fragment left over from the body's manufacturing of insulin. High C-peptide levels are associated with an astounding 270 percent[18] to 292 percent increased risk of subsequent colorectal cancer.[19] The women in the Nurses' Health

Study with the highest levels of C-peptide had a 76 percent higher risk of colon cancer.[20]

But excess insulin isn't an issue just for those with excess weight. Insulin levels form a spectrum. While those with obesity and diabetes have the highest insulin levels, healthy-weight nondiabetics may also have high insulin levels. Data from the National Health and Nutrition Examination Survey (NHANES) database from 1999 to 2010 suggest that high insulin levels more than double the risk of cancer *regardless of weight status*. Nonobese, nondiabetic participants with high insulin levels had a 250 percent increased risk of cancer deaths.[21] Hyperinsulinemia in normal weight women (BMI < 25 kg/m^2) doubles the risk of breast cancer.[22]

Injecting exogenous insulin, a drug more and more commonly prescribed for type 2 diabetes, also increases cancer risk. In the United Kingdom, the number of people with type 2 diabetes who are treated with insulin had grown dramatically from an estimated 37,000 in 1991 to 277,000 in 2010.[23] Weight gain is the major side effect, causing an estimated two kilograms' weight gain for every 1 percent reduction in HgbA1C, a blood test that reflects the average blood glucose over three months. This sounds ominous, as weight gain is a known risk factor for cancer. As researchers dug deeper, the news was not good. In the United Kingdom's General Practice Research Database 2000–2010, insulin treatment increased the risk of cancer by 44 percent compared to metformin, a treatment for type 2 diabetes that did not raise insulin levels. *Forty-four percent!* This was staggering. But this study was not the only one that found insulin treatment to be dangerous.[24]

Data from the province of Saskatchewan, Canada, confirmed that newly diagnosed diabetics starting treatment with insulin had a 90 percent higher risk of cancer compared to those taking the drug metformin.[25] The sulfonylurea drugs, which stimulate insulin secretion, were also linked to a 36 percent increased risk of cancer.[26] More insulin equals more cancer. It's a pretty simple concept. The longer one injects insulin, the higher the risk of cancer.[27]

We know that high insulin levels increase the risk of cancer. But why is insulin so important for cancer progression? Insulin is a hormone best known for its role in glucose metabolism. When we eat, insulin levels rise. When we don't, insulin levels decrease. Insulin is an important nutrient sensor, signaling the presence of food, but what does that have to do with cancer?

In a word: everything. The nutrient sensor insulin is also a highly potent growth factor.

16

GROWTH FACTORS

HEIGHT, OFTEN CONSIDERED a genetic trait, largely reflects the influence of growth factors during puberty. In post–World War II Japan, improved nutrition resulted in a gradual increase in height—and unexpectedly, breast cancer has risen in parallel.[1] Simply put, tall people get more cancer.[2]

Shockingly, of all the growth parameters commonly measured (birth weight, weight, height, age of menarche), the largest risk factor for breast cancer is height (see Figure 16.1).[3] In the United Kingdom, the Million Women Study found that the tallest women suffered 37 percent more cancer, particularly breast cancer.[4] Each increase of ten centimeters in height was associated with a 16 percent increase in risk of cancer. Growth factors increase height. Growth factors also increase cancer risk.

In addition to height, the rate of myopia, also known as nearsightedness, has been climbing for the last half century. Caused by eyeballs that have grown too long, myopia now affects over half of young adults in the United States and Europe, which is double the rate of fifty years ago.[5] A 1969 survey of an Alaskan village found only 2 of 131 people were nearsighted. But as their lifestyle changed, half their children and grandchildren were myopic.[6] The length of eyeballs in this village has been gradually increasing.

Look around. I wear glasses. I got teased mercilessly as a child in public school because . . . well, I was kind of nerdy. But more

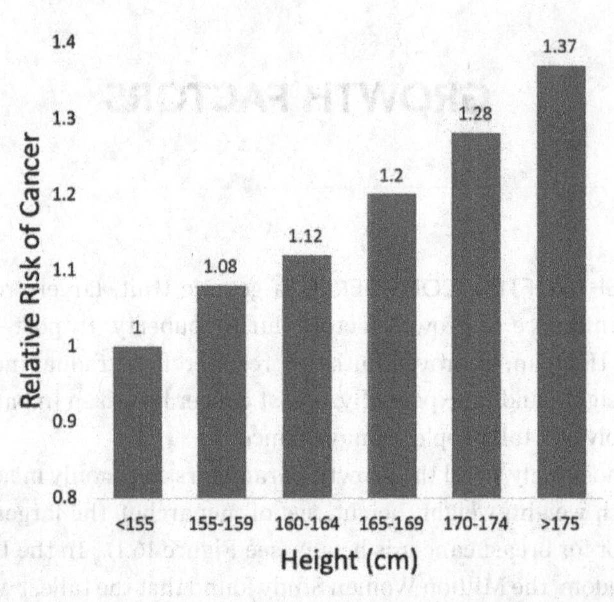

J. Green et al., "Height and Cancer Incidence in the Million Women Study: Prospective Cohort, and Meta-analysis of Prospective Studies of Height and Total Cancer Risk," *Lancet Oncology* 12, no. 8 (August 2011): 785–94, doi: 10.1016/S1470-2045(11)70154-1.

Figure 16.1: Increased risk of cancer with increased height.

than that, I was one of the very few kids in school wearing glasses. Looking around today's school rooms, I estimate that at least one third of the class wears glasses. Nobody gets teased, because everybody wears glasses. Nowadays, if boys didn't make passes at girls who wore glasses, there wouldn't be many girls to make passes at. The increasing rate of myopia cannot be primarily genetic, obviously, as it happened within a single generation. What's the link?

The common thread that runs through all the conditions of increased weight (obesity), increased height, increased eyeball length (myopia), and cancer is that they are all conditions of *excessive growth*. We often think that growth is good, but the truth is that in adults, growth is not necessary or even good. Quite the contrary. Growth is bad, sometimes very bad.

You don't want your eyeball to keep growing until it is the size of your head. You don't want your liver to keep growing until it squishes all the other abdominal organs out of the way. You don't want your fat cells to keep growing, because obesity causes many diseases, cancer being one of them.

Most of today's chronic diseases result from too much growth. The number one killer of Americans is cardiovascular disease, including heart attack and stroke. Excessive growth of atherosclerotic plaque blocks the blood vessels in the heart or brain, starving the tissue of life-giving oxygen. The number two killer of Americans is cancer, also a disease of excessive growth. Obesity is a disease of excessive growth. Fatty liver is a disease of too much growth. No, in adults, growth is *not* good. Further, the main determinant of this increased growth is not just genetics but growth factors—which brings us back to insulin.

INSULIN

Our understanding of insulin's surprising link to cancer began in 1985, with Dr. Lewis Cantley's discovery of the phosphoinositide 3-kinase (PI3K) pathway. Cantley, a professor at Harvard and Tufts and now director of the Meyer Cancer Center of the Weill Cornell Medical College, did not study metabolism or cancer, but the rather arcane field of cell signaling. The novel lipid molecule known as PI3K formed part of a previously unknown signaling pathway critical for regulation of cell growth.

What changed this discovery from a minor biochemical curiosity to a game-changing medical breakthrough was the key role that PI3K played in cancer. Experiments in the late 1980s found that cancer-causing viruses often activate PI3K[7] to levels one hundred times those seen in normal cells.[8] Unexpectedly, PI3K turned out to be one of the most significant oncogenes in human cancer. PI3K mutations rank fourth among the most common human cancer-causing gene mutations.[9]

High PI3K levels increase cell growth and promote cancer, so the next logical question is: what stimulates PI3K? As it turned out, the surprising answer was the well-known metabolic hormone insulin.[10] Something strange and unexpected was happening here. Insulin plays a huge role in metabolism (how cells generate energy), but was turning out to be a major regulator of cell growth, too.

From an evolutionary standpoint, this insulin/PI3K pathway is ancient,[11] having been conserved from worms and flies to humans. Almost all multicellular organisms use some variation of the insulin/PI3K pathway.

While today we think of insulin as a metabolic hormone, in primitive organisms, its primary function is as a growth hormone, regulating cell proliferation and survival. As we evolved into multicellular organisms, insulin evolved a second role as a nutrient sensor. This makes perfect sense, if you think about it, as growth always requires nutrients. When food is available, cells should grow. Make hay while the sun shines. When food is not available, cells should not grow. A multicellular organism growing too quickly when food is not available is a quick way to die.

When we eat, insulin and PI3K increase, which redirects the organism's priority toward growth. When we don't eat, insulin and PI3K fall, which redirects the organism's priority toward cellular repair and maintenance. PI3K provides this vital link between the nutrient-sensing and the growth pathways.[12] Insulin, in other words, stimulates cell growth, which has obvious implications for cancer.

Single-cell organisms such as yeast live directly in contact with their environment and therefore have little need for nutrient sensors. If food is available, the yeast grows. If food isn't available, it doesn't grow and instead forms dormant spores. So, yeast living on a slice of bread grows. Yeast living inside a plastic package doesn't grow. Upon exposure to water and sugar, the dormant spores reawaken and bloom. Survival for all life on earth depends upon this tight association between growth and nutrient availability.

Multicellularity means that some cells lose contact with the outside environment. Your kidney cells live all the way inside the abdomen, without any contact with the outside world. So, how do the kidneys know whether food is available? How do they know whether to grow or stop growing? Nutrient sensors evolved to translate outside food availability into hormonal signals. Now these nutrient sensors must be linked to growth signaling.

Using the exact same molecule (insulin) as both a growth factor and a nutrient sensor solves this fundamental coordination problem. When we eat, insulin goes up, which signals the availability of nutrients, and also signals the body to grow. Excessive nutrient sensing means excessive growth, a condition of obvious importance to cancer. When no food is available, insulin goes down, which also serves as the hormonal signal to stop growing. Reduced nutrient sensing means reduced cell growth. Growth and metabolism signals are one and the same.

So, how does this process work? As shown in Figure 16.2, insulin in the blood must first activate the insulin receptor on the cell surface. Many cancers carry too many copies of the insulin receptor, explaining the disproportionate effect of insulin on cancer growth. Insulin activates PI3K, which in turn activates, among others, two key pathways: a metabolic pathway and a growth pathway.

The well-known metabolic effects are mediated through the glucose transporter type 4 (GLUT4), which allows glucose to enter the cell and convert to energy. What Cantley discovered was the previously unrecognized importance of insulin/PI3K in stimulating cell growth. Insulin, through PI3K, activates the mTOR

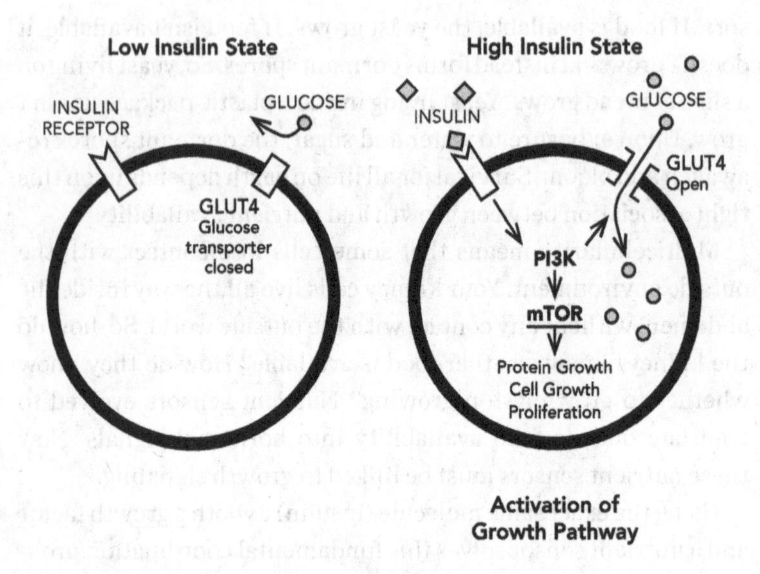

Figure 16.2

system (more on that in the next chapter), which stimulates cell growth and proliferation.

It is little wonder that cancer, a disease of excessive growth, loves the growth factor insulin. In the rare genetic disease called Cowden syndrome, a mutation along this pathway results in enhanced insulin signaling, substantially increasing the risk of both obesity and cancer.[13] The lifetime risk of cancer for those with this condition is a staggering 89 percent.[14]

The recognition that growth is intimately linked to nutrient signaling was a stunning revelation. It offered nothing short of a complete paradigm shift in our understanding of some human cancers, particularly the obesity-related cancers like breast and colorectal. Growth and nutrition/metabolism were now inextricably connected through the nutrient sensor insulin. A disease of excessive growth (cancer) is also *always* a disease of metabolism.

The nutrient sensor insulin is an important growth factor. Thus, hyperinsulinemia overstimulates the growth pathways and predisposes the body to diseases of excessive proliferation. Insulin also preferentially benefits cancer cells, which have higher glucose requirements because of the relatively inefficient energy-generating pathway of glycolysis. Obesity and type 2 diabetes, the prototypical diseases of hyperinsulinemia, significantly increase the risk of cancer. We had finally started to understand that aspect of diet that most influences cancer. Not fiber. Not dietary fat. Not vitamin deficiency. It was the nutrient sensor insulin.

Burkitt noted in 1973 that in Africa, the diseases of the West, including cancer, first appeared among the upper socioeconomic classes and in urban centers, where people had more access to imported, processed foods. Between 1860 and 1960, fat consumption increased by less than 50 percent, but sugar consumption more than doubled. The problem, he proposed, was that "The first change in traditional food is usually the addition of sugar. This is followed by substituting white bread for . . . cereals."[15] While the nutritional community was busy demonizing dietary fat, the scientific evidence pointed the finger directly at sugar and refined grains, consumption of which led to hyperinsulinemia. Dr. Lewis Cantley would say, decades later, "Sugar scares me."

Insulin-like Growth Factor 1

In a remote corner of Ecuador lives a community of about three hundred members known as the Laron dwarves, established in the fifteenth century by a group of Jews fleeing the persecution of the Spanish Inquisition. Geographic isolation produced inbreeding that led to the overexpression of rare genes—known in biology as the founder effect. In this case, the Laron dwarves are believed to have all descended from a single common ancestor, as they carry a shared rare mutation leading to short stature, or dwarfism. Their average height is only four feet, but they are otherwise typically formed.[16] Remarkably, this group appears to be completely immune to cancer!

Dwarfism is usually caused by low levels of growth hormone

(GH), which is responsible for the growth in height that typically occurs during puberty. GH stimulates the liver to secrete insulin-like growth hormone (IGF-1), which carries the growth message to the rest of the body. As the name suggests, IGF-1 and insulin are chemically very similar. The Laron dwarves had plenty of GH, but did not produce any IGF-1 due to a mutation of the GH receptor gene, and therefore suffered from short stature. But luckily for us, the story of the Laron dwarves did not end there.

By 1994, local researcher Jaime Guevara-Aguirre noticed that the Laron dwarves had a cancer rate of less than 1 percent, compared to 20 percent in their relatives who did not suffer from dwarfism. A more recent 2016 survey also found no incidence of cancer.[17] Interestingly, these patients were also protected from another dreaded disease: diabetes. Guevara-Aguirre could find obesity, but neither diabetes nor cancer.[18] Without the growth-promoting effects of IGF-1, cancer risk was substantially reduced.

So, other than GH, whose levels are typically quite stable, what else causes IGF-1 to go up? The answer, as you may have suspected, is insulin.[19] The signaling networks of both insulin and IGF-1 work through the PI3K pathway and are so closely intertwined that they are usually considered together in scientific publications. Too much insulin/IGF-1 means too much growth, which tilts us toward cancer, such as breast, endometrial,[20] prostate, and colorectal cancers.[21] In cell culture studies, adding IGF-1 increased colon cancer cell migration and metastasis—meaning it provided fertile soil for cancer to spread.[22] Increased levels of IGF-1 are associated with a 247 percent[23] to a 251 percent increased risk of developing colorectal cancer.[24]

But insulin/IGF-1 is not the only nutrient sensor in the human body. It is not even the most ancient one. That honor belongs to mTOR, the mechanistic target of rapamycin.

17

NUTRIENT SENSORS

THE STORY OF the nutrient sensor known as "mechanistic target of rapamycin," or mTOR, began in 1964, when microbiologist Georges Nógrády collected soil samples on the remote island of Rapa Nui, also known as Easter Island, and passed them to researcher Dr. Suren Sehgal, then with Ayerst Laboratories, for analysis. In 1972, Sehgal isolated the bacterium *Streptomyces hygroscopicus*, which produces a potent antifungal compound that he called rapamycin, after the island of its origin. He hoped to use it to make an antifungal cream that would treat athlete's foot, but this discovery turned out to have far, far more important implications.[1]

When Dr. Sehgal transferred labs, he wrapped some vials of rapamycin in heavy plastic, took them home, and stored them in his family freezer next to the ice cream. He was saving these specimens for a time when he could resume research on this fascinating new drug—which, due to other research priorities, would not be until 1987. When he did resume, he found that rapamycin proved to be a rather unremarkable antifungal, but that it had a potent suppressive effect on the immune system. Still, the mechanism of action was completely unknown.

By 1994, scientists discovered the protein that was the target of rapamycin and named it (imaginatively) the "mammalian target of rapamycin" (mTOR). The discovery of this protein then led to the discovery of a previously unknown and unsuspected biochemical pathway of human nutrition and metabolism.

This was an astounding revelation for biologists—the equivalent of suddenly discovering a new continent in the Atlantic Ocean. Hundreds of years of medical science had somehow missed this fundamental nutrient-sensing pathway that was so essential to life on earth that it had been conserved in animals from yeast all the way to humans. In an evolutionary sense, mTOR is older even than the much better-known nutrient sensor insulin. The mTOR pathway is found in virtually all life-forms, rather than just mammals, so the name was changed to *mechanistic* target of rapamycin, but it retained its catchy moniker "mTOR."

But what does it do? As shown in Figure 17.1, the mTOR pathway functions like a central command station by evaluating multiple sources of information before deciding whether to proceed with cell growth. mTOR considers several key sources of information, including dietary protein,[2] insulin, oxygen levels, and cellular stress. It is both a nutrient sensor itself (for protein) and an integrator of information provided by other nutrient sensors like

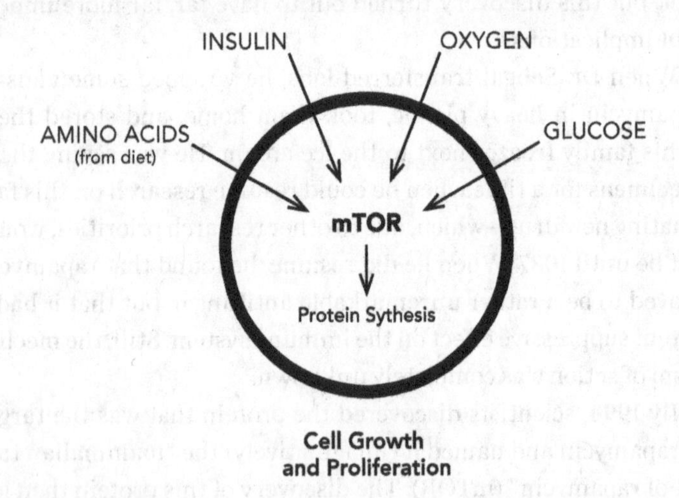

Figure 17.1

insulin. When mTOR becomes activated, it significantly increases cell growth.

The drug rapamycin blocks mTOR, which in turn stops cell growth, which explains how it functions as both an antifungal and an immune suppressant. The *Streptomyces hygroscopicus* bacterium secretes rapamycin to kill surrounding fungi by blocking their growth. Rapamycin also blocks human immune cell division, and therefore works as an immune suppressant.[3] By 1999, rapamycin was being routinely prescribed for liver and kidney transplant patients, to prevent organ rejection by their own immune systems.

Up until this point, most of the immune suppressants prescribed to transplant patients had the unfortunate side effect of increasing cancer risk—by a lot. According to the NIH, organ transplant recipients are at increased risk for thirty-two different types of cancer.[4] The immune system normally patrols the body looking to kill any stray cancer cells it finds. When the immune system is heavily suppressed to prevent organ rejection, cancer cells escape this immune surveillance.

But rapamycin was completely different. It suppressed the immune system but *decreased* the risk of cancer.[5] This was unprecedented! Rapamycin would eventually be shown to be effective against breast, prostate, and lung cancers. This was a major breakthrough in cancer treatment, introducing a whole new class of chemotherapy.

Rapamycin had unearthed a growth pathway previously unknown. The pathway mTOR is so deeply embedded in the growth decisions of normal human cells that aberrations of mTOR are estimated to occur in an astounding 70 percent of human cancers. Mutations in important cancer-causing genes such as PI3K, AKT, RAS, RAF, PTEN, NF1, and APC all work through their effect on mTOR.[6] When the nutrient sensor mTOR increases, cellular growth increases, as does the risk of cancer.

Insulin and mTOR are not the only nutrient sensors in the

human body. There is also the nutrient sensor known as AMP-activated protein kinase (AMPK).

AMPK

The nutrient sensors insulin and mTOR respond mainly to dietary intake of carbohydrates and proteins. The nutrient sensor AMPK, however, assesses the overall available cellular energy. When cells generate energy, regardless of the source (carbohydrates, proteins, or fats), AMP (adenosine monophosphate) is converted to adenosine triphosphate (ATP), which, as you may recall, stores potential energy. When energy is needed, ATP releases its energy and reverts to AMP.

When the cell has little energy available, it has lots of AMP and little ATP, and this stimulates AMPK. This nutrient sensor doesn't reflect what you just ate, but acts more like an overall fuel gauge of how much energy the cell has remaining.

- Lots of energy stored = low AMPK.
- Little energy stored = high AMPK.

Just like the other nutrient sensors, mTOR and insulin, AMPK is also inextricably linked to growth. High AMPK lowers mTOR activity, slowing down growth. AMPK increases production of new mitochondria, the energy makers in cells, to increase the cell's capacity for burning fat. AMPK also increases autophagy, the important cellular self-cleansing and recycling process.

Drugs that activate AMPK (mimicking low cellular energy stores) are known for promoting health. Examples include the diabetes drug metformin; resveratrol, from grapes and red wine; epigallocatechin gallate (EGCG), from green tea and chocolate; capsaicin from peppers; turmeric, garlic, and the traditional Chinese medical herb berberine. Calorie restriction also activates AMPK, which may explain some of its purported benefits with regard to aging.

NUTRIENT SENSORS

Most omnivores, like humans, eat when food is available. Traditionally, we have enjoyed a wide bounty of crops in the summer and fall—food is plentiful, and we are able to consume a lot of energy. But before the advent of grocery stores, when people were living off the land, little food was available in colder months. Humans survived these times of scarcity because we have well-developed systems for energy storage (body fat) and also highly conserved nutrient sensors that signal our cells to grow when food is available and not to grow when food is not available.

The three most important nutrient-sensing pathways in humans are insulin, mTOR, and AMPK. Each nutrient-sensing pathway provides different yet complementary information (see Figure 17.2). Insulin increases primarily in response to dietary carbohydrates and protein, and responds within minutes. mTOR increases primarily in response to dietary protein and plays out over eighteen to thirty hours. AMPK responds to overall cellular energy, which reflects the intake of all macronutrients. Its total effect runs over the longer term, from days to weeks.

Nutrient Sensor	Macronutrient	Time Scale
Insulin	Carbohydrates, protein,	Short term
mTOR	Protein	Medium term
AMPK	Carbohydrate, protein, fat	Long term

Figure 17.2

By using three different nutrient sensors, cells gain exquisite information about the type of food available and also the duration of its availability. Are nutrients mostly fat or carbohydrate or protein? Are nutrients available only temporarily or long term? Crafted by millions of years of evolution, the biochemical wizardry of our nutrient sensors makes a mockery of our comparatively dull caveman brain that can only say, "Look like food to Grok. Grok eat."

All three nutrient sensors are interconnected and directly linked to cellular proliferation. When nutrients are available, cells grow. An environment full of nutrients and growth signals provides fertile soil for cancer. When nutrients are not available, cells don't grow. But when food is scarce, it's not enough for the total cell population to simply stop growing; it must actively shrink. Cells and subcellular parts must be culled in the processes known as apoptosis and autophagy, respectively. If these important systems go awry, the odds may tilt toward excessive growth and cancer.

APOPTOSIS

Growth is essentially the balance between two opposing forces: the rate of cellular growth and the rate of cellular death. Overall growth occurs when too many cells are proliferating or too few cells are dying. Insulin/IGF-1 promotes cell proliferation, but plays an equally large role in preventing cell death, or apoptosis (see chapter 3). Species survival depends upon matching growth to nutrient availability. Too few nutrients for too many cells equals death. When nutrients are scarce, the logical action is to remove some extraneous cells. Like a freeloading uncle who has overstayed his welcome, these extra mouths must go.

The single-cell organisms' prime directive is to survive and multiply at all costs. If their death is messy . . . well, that's a mess for somebody else to worry about. However, multicellular organisms require careful coordination for both adding and deleting cells. Dying and dead cells harm the cells around them—one bad apple spoils the whole barrel. In multicellular organisms, a controlled method of removing that bad apple is essential.

A normal human body makes approximately ten billion new cells each and every day. This also means that ten billion cells must also die and be neatly removed each and every day.[7] Apoptosis removes these cells in a regulated and nontoxic manner. The cell carefully marked for removal undergoes controlled changes and eventually splits into small pieces (apoptotic bodies) that are safely disposed of. Cell contents do not randomly splatter out, as in necrosis.

Apoptosis must be tightly regulated, as either too much or too little is pathologic. If a cell resists apoptosis (a hallmark of cancer), the delicate balance is tipped to favor growth, which favors cancer. Apoptosis thus forms a key defense mechanism against damaged or dangerous (cancerous) cells. So, how is it controlled?

Two main pathways activate apoptosis: the extrinsic pathway (also called the death receptor pathway) and the intrinsic pathway (also called the mitochondrial pathway). Once triggered, apoptosis cannot be stopped. For the purposes of this discussion on cancer, I will focus on the mitochondrial pathway, which is controlled by both positive and negative stimuli. Positive stimuli are factors that initiate apoptosis, including cellular damage from toxins, viruses, radiation, heat, and insufficient oxygen. The body does not want broken cells all over the place, so apoptosis efficiently clears damaged cells without the accompanying mess. It is no coincidence that these positive stimuli for apoptosis are also carcinogens. Damaged cells are supposed to die. If they do not, they may become cancerous.

Negative stimuli are default pathways that automatically trigger unless appropriate signals are present. For example, you can sign up for a free trial of Amazon Prime, and if you don't contact them in time at the end of the free-trial period, you will be automatically enrolled. Apoptosis works the same way. If no signal (growth factor) to stop apoptosis is received, the cell involuntarily starts apoptosis. This dual-control structure, using both positive and negative stimuli, is far more robust and makes apoptosis a particularly effective anticancer strategy.

Which growth factors prevent apoptosis? The best-studied

anti-apoptotic factor is insulin/IGF-1, through the PI3K pathway.[8] High levels of insulin/IGF-1, as seen in obesity and type 2 diabetes, not only encourage cell growth but also block the natural running of the apoptosis program, powerfully increasing growth signaling. High insulin/IGF-1 thus forms part of a fertile soil for the cancer seed.

In cancer, the balance between proliferation and destruction is tilted fatally toward proliferation. Defects in apoptotic pathways allow damaged cells to survive when they should have been marked for death. But proper execution of the mitochondrial pathway of apoptosis depends upon one key organelle: the mitochondria. If the mitochondria are dysfunctional, the mitochondrial pathway of apoptosis will similarly not work, tilting the balance toward growth and cancer.

MITOCHONDRIA

In our distant evolutionary past, mitochondria existed as separate organisms. They were engulfed by a primitive cell approximately two to three billion years ago and developed a mutually beneficial arrangement. The cell provided shelter and nutrients to the mitochondrion, and in return, the mitochondrion performed various tasks, including energy generation and eventually apoptosis.

Mitochondria are highly susceptible to damage and, to stay healthy, are constantly remodeling. To maintain high-quality mitochondria capable of carrying out apoptosis, two things must happen: old or damaged mitochondria are removed through a process called mitophagy, and new mitochondria are created.

Mitophagy is closely related to the cellular process called autophagy, which was brought to prominence by Nobel laureate Dr. Yoshinori Ohsumi. The word *autophagy* derives from the Greek words *auto*, meaning "self," and *phagein*, meaning "to eat," so it literally means "eating oneself." Autophagy is a regulated, orderly process of degrading cellular components to be recycled into new ones. Autophagy functions as a cellular housekeeper and is controlled mainly through the nutrient sensor mTOR. When lots of nutrients are available, mTOR is high, putting the cell into "growth" mode, so autophagy and mitophagy turn off. As always, the processes of cell growth/degradation and nutrient availability are indivisible. Without mitophagy and the removal of old mitochondria, new ones cannot be formed.

The major signal to generate new mitochondria is the nutrient sensor AMPK.[9] When overall energy availability is low, AMPK is high, which stimulates the growth of new mitochondria. In animal models, AMPK restriction and food restriction have been shown to maintain healthy mitochondrial networks and increase life span.[10] Animals exposed to intermittent fasting showed a striking benefit in mitochondrial networks.

On the one hand, excessive nutrient availability, detected through the nutrient sensors insulin, mTOR, and AMPK, reduces mitophagy and new mitochondrial formation. To maintain healthy mitochondria, you don't need *more* nutrients, but you periodically need *fewer* nutrients. Defective mitochondria impair apoptosis, which swings the delicate balance between cell growth and cell death. Damaged cells allowed to persist may turn cancerous because of the selection pressure to survive. These cells should have been culled but were not. The periodic removal of older or damaged cells forms one of our primary anticancer defenses.

On the other hand, nutrient deprivation—especially protein deprivation—lowers mTOR and activates autophagy. This takes the cell out of growth mode and into a maintenance/repair mode. Old, defective cells and organelles are culled. If there are not enough nutrients, then the cell does not want to maintain all these extra parts. When nutrients become available, autophagy turns off, putting the cell back into growth mode.

Any combination of increased growth factors or decreased cell death (apoptosis) will allow progression of a cancerous cell toward growth. Recent research points out that growth factors and nutrient sensors are indivisible from one another. Nutrient sensors are growth factors. Therefore, diseases of growth are also always diseases of metabolism, explaining the singular importance of insulin in facilitating cancer. Perhaps not coincidentally, the mitochondria are the key site for both energy metabolism and apoptosis.

High blood insulin (hyperinsulinemia) causes the metabolic diseases of obesity and type 2 diabetes and, through PI3K and IGF-1, facilitates a disease of growth cancer. But the idea that cellular metabolism might be important in cancer is not new. Over a century ago, one of history's greatest biochemists, Nobel laureate Otto von Warburg, proposed that the key to understanding the origins of cancer was to look at its metabolism.

PART V

METASTASIS

(Cancer Paradigm 3.0)

18

THE WARBURG REVIVAL

NOBEL LAUREATE OTTO Heinrich Warburg (1883–1970) was born in Freiburg, in southwestern Germany. He was the son of Emil Warburg, a prominent physics professor at the University of Freiburg, and grew up around contemporaries like Albert Einstein and Max Planck, both of whom would also become legendary scientists.

Otto Warburg's specific research interest was cellular energetics, where he applied the rigorous methods of the physical sciences (chemistry and physics) to biology. How much energy do cells require? How do they generate that energy? This obsession would lead him eventually to his life's research: what he called the "cancer problem." How did cancer cells differ in their energy metabolism from normal cells?

Normally, cells may generate energy in the form of adenosine triphosphate (ATP) in two different ways: oxidative phosphorylation (OxPhos), also called respiration; and glycolysis, also called fermentation. OxPhos, which occurs in the mitochondrion, burns each glucose molecule with oxygen to generate thirty-six ATP. Without oxygen, normal cells must resort to glycolysis, which generates only two ATP and two lactic acid molecules per glucose molecule. For example, during intense exercise, muscles require so much energy so quickly that the blood flow cannot keep up with the oxygen demand. The muscles switch to glycolysis, which does not require oxygen and generates far less energy per glucose molecule. Eventually, lactic acid builds up, causing muscle fatigue, which is why you may feel like you can't go any further in the midst of a particularly challenging workout. Normal cells do not function well in an acidic environment. As your body rests, oxygen demand slows, and your muscle cells resume OxPhos once oxygen supply catches up.

Cells generating a lot of energy using OxPhos naturally require more oxygen. Warburg noted this phenomenon when he observed fertilized sea urchin eggs growing rapidly. He speculated that rapidly growing cancer cells would also consume oxygen prodigiously. But he was wrong. In 1923, Warburg noticed with some amazement that fast-growing rat tumor cells used no more oxygen than regular cells.

The cancer cells were instead using ten times more glucose and producing lactic acid at seventy times the rate of normal tissues

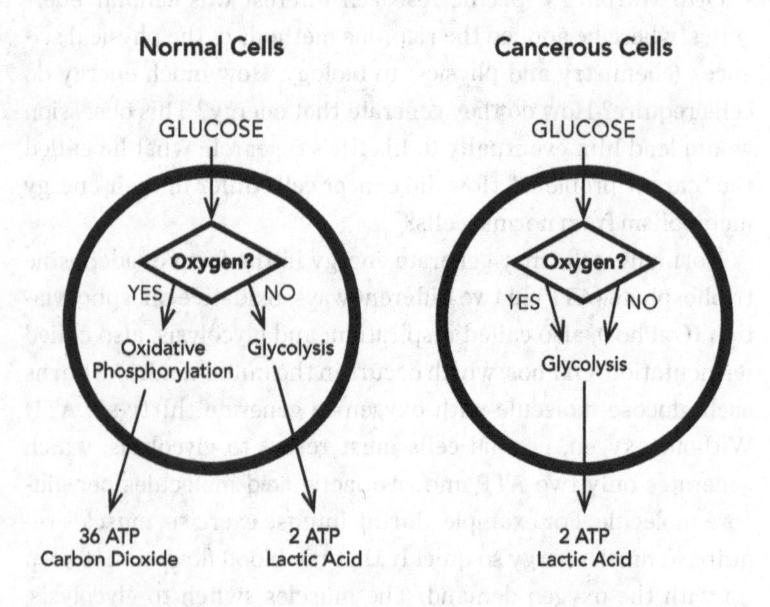

Figure 18.1

(see Figure 18.1).[1] Warburg calculated that tumor cells converted an astounding 66 percent of the glucose they took up to lactate.[2]

Despite the ready availability of oxygen, cancer cells were generating energy using the less efficient glycolysis pathway. This surprising process is now referred to as the Warburg effect.

Because glycolysis generates much less ATP per glucose, cancer cells must drink up glucose like a camel drinks water after a long desert trek. Today we use the Warburg effect for the common cancer imaging test, the PET scan. As we discussed in chapter 3, PET scans measure how much glucose is being consumed by cells. Active cancer cells guzzle the glucose much faster than normal surrounding cells, and the PET scan picks up these hot spots.

Aerobic glycolysis (glycolysis in the presence of lots of oxygen) is unique to cancer. Normal cells almost always choose OxPhos if oxygen is plentiful. Even in situations where cells grow quickly and require large amounts of energy, such as during wound healing, the Warburg effect is not found. But why? It seems very strange.

Think about it. We know that cancer can be distinguished by four hallmarks:

1. It grows.
2. It is immortal.
3. It moves around.
4. It uses the Warburg effect: it deliberately uses a less efficient method of energy extraction.

One of these does not fit with the others. Immortal cancer cells are super busy, constantly growing, and moving all over the body. This requires lots and lots of energy. Why on earth would cancer choose a *less* efficient manner of energy extraction?

Suppose you build a fast sports car, sleek and low to the ground and with a spoiler on the back to reduce drag. Then you take out its 600-horsepower motor and put in a 9-horsepower lawnmower engine. Huh? That's bizarre. Cancer effectively does the same thing by deliberately choosing a less efficient energy-generating method. Yet it cannot simply be a coincidence, as about 80 percent of all known cancers use the Warburg effect. Whatever the reason, it is critical to cancer's genesis and not merely a metabolic mistake. Cancer has not persisted for millennia, in animals ranging from hydra to dogs to cats to humans, by making mistakes.

In a famous 1956 research paper titled "On the Origin of Cancer," Warburg hypothesized that the anomalous switch to aerobic glycolysis is so bizarre that it must be cancer's inciting event. To review, the two main requirements for OxPhos are oxygen and functional mitochondria, the cellular structures where OxPhos takes place. Because oxygen is plentiful, Warburg deduced that it must be the mitochondria that are dysfunctional, forcing the cancer cell to revert to the less efficient glycolysis pathway.[3] Warburg hypothesized that cancer was caused primarily by mitochondrial damage.

While the Warburg effect is a well-established fact, many observations argue against Warburg's hypothesis.[4] Mitochondria from cancer cells often function normally with preserved respiration.[5] Most cancer cells have normal mitochondrial function, meaning they aren't exclusively reliant on glycolysis for energy production—they could switch back to OxPhos if necessary.[6] Cancer was not being *forced* to use glycolysis, it was *choosing* it. But why?

Efficiently generating energy (ATP) is an advantage only under conditions of scarcity. If there is a lot of glucose around, then why does it matter if each glucose produces only two ATP instead of thirty-six? Glycolysis produces ATP less efficiently but more quickly. In the time that normal cells metabolize one glucose to thirty-six ATP, cancer cells metabolize eleven glucose molecules to twenty-two ATP and twenty-two lactic acid. Because lactic acid can be converted to ATP one to one, this gives cancer a potential total of forty-four ATP. Cancer cells produce energy quicker, although it requires ten times more glucose to do so.[7]

Imagine two people. One burns 2,000 calories per day while the other, being more energy efficient, burns only 1,000 calories. The increased energy efficiency is not an advantage if you are eating 2,500 calories per day. OxPhos is advantageous only when glucose is scarce, but given the recent obesity and type 2 diabetes epidemics, glucose levels tend to run high, not low. So, the OxPhos "advantage" of energy efficiency is largely illusory in this current environment.

The fact that almost every known cancer uses this pathway suggests that it is neither coincidence nor a mistake, but integral to cancer development. It must confer some selective advantage. But what?

Cells need more than just energy to grow. They also need basic building blocks. Because we are carbon-based life-forms, our cell growth is reliant on carbon to build basic molecules. During OxPhos, most of the carbons in glucose are metabolized for energy, leaving behind carbon dioxide, which is exhaled. During glycolysis, only a small percentage of carbons are completely burned for energy. The leftover carbons can be metabolized into carbon building blocks to make new amino acids and fatty acids.

Consider this analogy: Building a house requires both energy (the hard work of the builders) and materials (bricks). Having builders but no bricks is useless. Similarly, rapidly growing cells require both energy (ATP) and materials (carbons). OxPhos generates pure energy alone, which does not maximize growth. Glycolysis better supports rapid growth because it provides *both* energy and materials, while OxPhos generates only pure energy.[8] This may explain the advantage of the Warburg effect for cancer growth.

By the 1970s, Warburg's focus on cancer cell metabolism was looking increasingly shaky. The genetic revolution was well under way, and cancer researchers were drawn to the somatic mutation theory like iron filings to a magnet. The question of how cancer fueled its growth and the anomalous, curious predilection for glycolysis was a conveniently ignored mystery. Entire years would pass without any scientific publications on the Warburg effect. These two scientific fields of inquiry, cancer's growth and cancer's metabolism, were complete strangers to each other. Then, unexpectedly in the late 1990s, they united in a shotgun marriage.

THE WARBURG REVIVAL

The pathways that govern a cell's growth and its metabolism had always been considered distinct. But Lew Cantley's groundbreaking research linked the well-known metabolic hormone insulin directly to growth pathways through PI3K. Cancer cell growth and metabolism were inextricably linked by the exact same genes and hormones.[9] For example, the oncogene *myc* controls not only growth, but also a metabolic enzyme that turns on the Warburg effect. Cantley found a direct link between nutrient sensors, metabolism, the Warburg effect, and cellular proliferation.[10] Genes controlling growth also controlled metabolism.

All these newly discovered oncogenes and tumor suppressor genes also influenced metabolic pathways. Many oncogenes control enzymes called tyrosine kinases, which regulate both cell growth and glucose metabolism. The ubiquitous *p53* tumor suppressor gene influences growth, and it also regulates cell metabolism by affecting mitochondrial respiration and glycolysis.

Cancer cells can't stop growing, but they also can't stop eating. Does cancer grow because it can't stop eating, or does it eat because it can't stop growing? Most likely, both. Diseases of growth *were* diseases of metabolism—and this applied to more than just glucose metabolism.

Cancer cells love eating glucose, but not exclusively. The metabolic pathways of the amino acid glutamine are also disrupted in cancer.[11] Amino acids are the building blocks of proteins, and glutamine is the most abundant amino acid in the blood. Some cancer cells consumed more than ten times the normal amount of glutamine.[12] Some cancers, such as neuroblastoma, lymphoma, and kidney and pancreatic cancer, appeared so "addicted" to glutamine that they simply couldn't survive without it.[13]

Warburg believed that cancer was solely dependent on glucose for energy, but this was not entirely true. Cancer can also metabolize glutamine, and more recent studies show that cancer may also metabolize fatty acids and other amino acids.[14] Cancer competes with other cells for fuel in a crowded environment, so having the flexibility to use a variety of fuels is advantageous for growth. While Warburg's original hypothesis may not have panned out, his hunch that cancer's metabolism was vitally important was spot on. The Warburg effect *did* have a purpose. It provided cancer cells with a strategic advantage in their struggle for survival. The large volume of lactic acid produced during the Warburg effect was not a waste product, as previously assumed, but a major benefit, providing the cancer cell with a significant survival advantage.

LACTIC ACID

As a tumor grows, new cancer cells crop up farther and farther away from the main blood supply that provides oxygen and clears waste products. Cells closer to the blood vessels are well supplied and thrive. Cells farther away do not get enough oxygen to survive. In between these two regions is the area known as the hypoxic zone, where cells receiving barely enough oxygen to survive activate an enzyme called hypoxia-inducible factor (HIF1). The struggle for survival in this hypoxic zone acts as a potent selective evolutionary pressure.

First, HIF stimulates the release of vascular endothelial growth factor (VEGF), which promotes the growth of new blood vessels. New blood reserves deliver more oxygen and allow the tumor to grow larger. "Inducing angiogenesis" is one of the key hallmarks of cancer described by Weinberg and Hanahan.

Second, HIF makes it easier for the normally stationary cells to become more mobile. The adhesion molecules that anchor cells to their proper position are disrupted, and the basement membranes that limit cells to certain areas are degraded.[15] This makes it easier for cells to "activate invasion and metastasis," another key hallmark of cancer.

Third, because oxygen is scarce, HIF reprograms the cell's metabolism from OxPhos and toward glycolysis. Because more glucose is needed for energy production, HIF increases the expression of cellular glucose receptors. At the same time, HIF decreases the production of new mitochondria, which are essential for OxPhos.[16] In essence, HIF is responsible for the phenomenon known as the Warburg effect, yet another key hallmark.[17]

This package of changes induced by HIF improves survival in a low-oxygen environment. Cells deprived of oxygen attempt to build new blood vessels, move away from the hypoxic region, and use less oxygen. Not coincidently, these are also behaviors typical of cancer cells, and this is precisely *the environment that provides a unicellular organism with an advantage over that of its multicellular cousin*. The Warburg effect is not simply a metabolic "mistake." It provides cancer cells with a unique survival advantage when competing against other cells.

Cancer cells produce lactic acid during glycolysis, and they dump that acid into their surrounding environment in the same way a chemical plant might dump toxic waste in its surroundings. This is no accident, and the lactic acid is not merely a waste byproduct. Tumors are using precious energy to deliberately manufacture and pump more acid into their immediate surroundings, which are already acidic.[18] Compared to normal cells, which live in an environment with a pH of 7.2 to 7.4, tumors generate a surrounding microenvironment of pH 6.5 to 6.9.[19] Why do cancer cells put so much effort into acidifying their surroundings?[20] Because the acidity give the cells a huge survival advantage. Normal cells are injured in an acidic environment and undergo apoptosis, while cancer cells tolerate the acidity fairly well.

There are two ways to win: get better or make your competitor worse. Both work. Cancer is playing a cellular Game of Thrones. You win or you die. While normal cells play nice and cooperate, unicellular organisms compete by sabotaging opponents. Cancer cells secrete the noxious lactic acid to impede nearby cells. Killing your neighbors is a time-tested survival strategy, and is common in the unicellular universe.

In 1928, Sir Alexander Fleming discovered that the fungus *Penicillium notatum* secreted a noxious chemical into its surroundings that killed competing bacteria. That chemical eventually became the breakthrough antibiotic penicillin. On Easter Island, rapamycin was discovered from a bacterium that secreted a noxious chemical into its surroundings to kill competing fungi.

The caustic acidic environment degrades the extracellular matrix, the normal supporting structure of the cell. This allows for the cancer cell to invade through the basement membrane more easily, an important prerequisite for metastasis. The damage caused by the lactic acid also provokes inflammation. This attracts immune cells that secrete growth factors, which would be useful in wound healing, but ultimately benefit the cancer cell.

Cancer has been called "the wound that never heals" because of its similarity to the hypergrowth seen in wound healing. In normal wound healing, new blood vessels replace the torn old ones, cellular debris is cleared, and the wound heals. The main difference is that the wound-healing program eventually comes to an end, whereas cancer's growth program does not.

Even when oxygen is freely available, cancer continues to use glycolysis because it offers the singular survival advantage of pumping out lactic acid (the Warburg effect). The inflammation induced by lactic acid also inhibits those immune cells that normally target and kill cancer cells.[21] Thus, the increased lactic acid from the Warburg effect:

- Suppresses normal cell function;
- Degrades the extracellular matrix, facilitating invasion;
- Provokes inflammatory response and growth factor secretion;
- Reduces immune response; and
- Increases angiogenesis.

Cancer didn't choose glycolysis over OxPhos (the Warburg effect) by accident. It's not a mistake. It's a logical choice because of the survival advantage offered by lactic acid. The tradeoff is the requirement for much more glucose as feedstock. During conditions of ample glucose, the balance is tipped toward cancer growth. It is the Warburg effect that sets the stage for the next step in the development of cancer by making it easier for cells to invade tissues and move around. This stage is largely responsible for cancer's lethality.

19

INVASION AND METASTASIS

THERE IS NO word in the cancer lexicon more terrifying than *metastasis*. The National Cancer Institute defines metastasis as "The spread of cancer cells from the place where they first formed to another part of the body."[1] This single distinguishing feature makes cancer more lethal than virtually any other disease in existence. A fact highlighting the severity of this phenomenon: metastases are responsible for an estimated 90 percent of cancer deaths.[2]

Like cancer cells, infectious diseases may also metastasize. Bacteria from a urine infection may spread to the kidneys, then to the blood, and settle into the heart valves. Bacteria move constantly but are not inherently malicious—they are simply organisms looking out for their own survival. Metastasis, or movement of cells, is an innate feature of single-cell life on earth.

Cancers are categorized as either benign or malignant. Both types of cancer behave identically in all respects, except that benign cancers lack this metastatic ability and therefore cause almost no significant disease. For example, a benign tumor of fat cells, called a lipoma, is very common, affecting an estimated 2 percent of the population. Mostly harmless, these tumors can grow to massive proportions. In 1894, a lipoma was removed weighing an estimated 50 pounds (22.7 kilograms).[3] Despite their frequency and sometimes enormous size, they are generally no deadlier than acne.

By contrast, malignant cancers are categorized as such because they can metastasize. If breast cancer stayed within the breast, then it would be easy to treat: simply cut it off. Once breast cancer cells have disseminated throughout the body, it becomes a highly lethal disease. So how does this metastatic process happen?

The metastatic cascade occurs in two steps: invasion of surrounding tissue and metastasis to distant sites. First, cancer cells must break free from the original tumor mass, which may occur even if the tumor is tiny. In the disease known as carcinoma of unknown primary, the metastatic cancer is detected without a primary site being found, because either it is too small or it has already disappeared. Most cells in multicellular species have adhesion molecules that anchor cells to their proper homes. Cancer cells must overcome these anchors to roam free.

Second, the cancer must invade through the basement membrane of the normal tissues. Just as potting soil sold in a plastic bag keeps your car nice and tidy, all cells are constrained by a membrane that keeps normal cells where they belong: inside the tissue of origin. To spread to other tissues, cancerous cells must burst through this basement membrane that keeps it wrapped up.

Once through, it can spread to surrounding tissues, local lymph nodes, or past blood vessel walls in a process called intravasation. Once in the bloodstream, cancer can travel along this highway to distant sites.

INVASION

1. Primary tumor formation.
2. Local invasion.
3. Intravasation.

METASTASIS

1. Survival in circulation.
2. Extravasation.
3. Survival at distal site.

These first three processes (primary tumor formation, local invasion, and intravasation) can be considered together as the process of invasion. The hypoxic, acidic environment around the cancer created by the Warburg effect clears a path for invasion.[4] Believe it or not, this is the easy part for cancer cells. Most cancers are able to accomplish this feat, given enough time. In experimental animal models, cancers escape to the bloodstream easily, almost 80 percent of the time. But as for a teenager bound for college, it's not leaving one's parents' house that's difficult; it's thriving on one's own that's the challenge. To metastasize, the cancer cells must survive the journey in the bloodstream, leave the blood vessel to invade a foreign organ, and then learn how to survive and thrive in that new environment.

Once cancer cells enter the bloodstream, the attrition rate is much, much higher, and metastasis is an order of magnitude more difficult to achieve. The bloodstream is a harrowing, hostile environment, with a million ways for cancer cells to die. Natural killer cells of the innate immune system hunt down and attack them immediately. This part of the immune system is called "innate" because the killer cells are naturally programmed to attack cancer on sight. The turbulence of the bloodstream is also an ever-present danger to cancer cells. Cells are generally stationary and lack the ability to handle the sheer force of a rushing torrent of blood. Many cancer cells are simply ripped apart in the current.

If the cancer cell somehow survives the traumatic journey and arrives on the distant shores of a foreign organ, it must leave the bloodstream and enter the organ, a process called extravasation. This sounds a lot easier than it actually is. The bloodstream is constantly moving, so attaching onto the wall of the blood vessel is no simple matter. Imagine being swept away by a raging river rapid and trying to latch yourself onto the shore using only your pinkie finger. The river is constantly pummeling you, threatening to sweep you back in. The cancer cell must somehow grab ahold of the edge of a smooth blood vessel, hang on against the rapidly flowing blood, and then exit through the blood vessel wall into the new organ.

The cancer cell now faces the hostile environment of an alien organ for which it is fantastically ill-equipped. For example, a breast cancer cell landing in the lung is completely bewildered by the alien environment. What is all this air blowing in and out? Where are the milk ducts? It's akin to a penguin that normally lives in the frigid climate of Antarctica suddenly showing up in the Sahara Desert.

Even after establishment in the new location, the new cancer immigrants must still thrive and proliferate—no easy task against the hostile incumbent cells. The entire time, the immune system is still doing its best to kill the cancer. Many such cancer cell colonies, called micrometastases, may persist for long periods of time, unable to grow, but strong enough to hang on.[5] Then the cancer cell must learn not just how to survive, but to proliferate and grow. This metastatic process requires extraordinary new survival skills utterly different from anything the cell has done before. Clearly, a cancer cell must have radically transformed its genetic makeup. So how do these changes occur?

Classically, the somatic mutation theory envisions a relatively orderly disease process, starting with a single cancer cell that has randomly accumulated several of the right mutations. The cancer grows larger, like a drop of red wine on a white tablecloth. When it becomes big enough, cancer cells break off into the blood. Some lodge in a distant organ, like the liver, and grow. It used to be widely believed that all these genetic mutations required for invasion and metastasis were accumulated by chance, but this "random accumulation of genetic mutations" hypothesis is now known to be incorrect.

No single "metastasis" gene has ever been discovered, despite decades of research and many thousands of genome-wide sequencing studies. Metastasis has defied all the genetic research of the last half century. That's because instead of a single gene, it takes the coordinated and precise mutation of hundreds of genes to successfully metastasize.

So, why would a cancer want to develop the hundreds of mutations needed to survive in hostile alien environments before it even left its home? It's like mortgaging your house to buy expensive equipment and training to survive on Saturn. With no current plans to establish a human settlement on Saturn, it is a huge waste of money and time. Why would the cancer direct massive resources toward metastasis to the liver or lung or bone *before* it spread? It is not a random process of accumulation, but instead, an evolutionary one. That is, cancer does not randomly accumulate the ability to invade and metastasize; it *evolves* that ability.

CIRCULATING TUMOR CELLS AND MICROMETASTASIS

Metastasis is an extraordinarily inefficient process. Given the almost insurmountable problems of metastasis, the majority of cancer cells sent forth from the primary tumor will die. Cancer cells reproduce every one to two days, but the doubling time of tumors is sixty to two hundred days, implying that the vast majority of cancer cells actually don't survive.[6]

So, how does the cancer overcome these hurdles? Once again, the answer can be found through the application of evolutionary biology to the cancer problem. Cancer does not follow an orderly progression of growth, invasion, and metastasis. Recent research indicates that metastasis, shockingly, is not a late phenomenon of cancer. It is actually one of the *earliest* steps cancer takes.

If cancer moved stepwise from growth to invasion to metastasis, then early extensive local excision at any time before metastasis would be curative. But the failure of the "radical" cancer surgeries performed in the first half of the twentieth century argues against this paradigm. Many microscopic and otherwise undetectable cancer cells (micrometastases) would have escaped long before clinical detection and surgery.

In cases of "cancer of unknown primary," which constitutes approximately 5 percent of cancer cases,[7] a widely metastatic cancer is found, but no primary tumor can be identified despite intensive investigations and imaging. Even at *autopsy*, 20 to 30 percent of cases remain unsolved. The primary cancer is so small as to be undetectable by all our modern technology, but somehow it still metastasized. This is largely because metastasis is an early, not a late, step in oncogenesis.

Recent technological advances have allowed us to detect cancer cells in the bloodstream, called circulating tumor cells (CTCs), even at extremely low concentrations. The discovery of these short-lived cancer cells in the blood has revolutionized our understanding of the metastatic cascade. The primary tumor sheds cancer cells into the blood from a very early stage, often when the primary tumor is itself undetectable. In the bloodstream, CTCs don't survive long. It is estimated that most live only a few hours,[8] which is why they were not recognized until recently. Almost all CTCs are destroyed as soon as they are released. Like the first wave of soldiers bravely storming the beaches of Normandy, the cancer cells are immediately annihilated by the body's formidable anticancer defenses.

Large numbers of CTCs are continually breaking off from the primary tumor and are destroyed quickly in the bloodstream, which is why metastatic cancer is rarely detectable at this early stage.[9] It is extremely difficult for CTCs to establish permanent colonies, even with millions of cancer cells flooding into the bloodstream every day. Most CTCs are simply killed off[10]—but not always.

Micrometastasis explains the phenomenon of cancer of unknown primary. CTCs leave the primary tumor very early and, for some unknown reason, are more successful in their new environment than the old environment. While the primary tumor may be small or destroyed entirely at its original location, the metastatic cancer flourishes, having found fertile soil. Thus, the metastatic lesion is detected before the primary tumor, which is sometimes never found.

Early metastatic cells may lodge in protected niches for many years, hiding from the anticancer forces. For example, breast cancer patients with known micrometastases had only a 50 percent chance of developing clinically detectable metastases within ten years.[11] The breast cancer had already spread its deadly seed, but it could not grow without fertile soil.

In the phenomenon of latent metastasis, cancer patients who have been "cured" occasionally relapse with distant metastases years or even decades after the cancer was believed to be long gone. The body's anticancer defenses were able to keep the micrometastatic foci in check for a while, but over time, the cancer cells gained a foothold.

Early metastasis also explains the need for local radiation and chemotherapy after surgery. During surgery, all the visible cancer is removed. Nevertheless, most cancer protocols still require postoperative radiation or chemotherapy. If cancer was indeed an orderly process, then these other measures would not be necessary. Radical surgery would have removed all traces of cancer. But because cancer metastasizes early, we still need these additional treatments.

TUMORAL EVOLUTION AND SELF-SEEDING

Considering the complexity of metastasis, the classical SMT view of "random accumulation" of the necessary hundreds of mutations working together is barely in the realm of possibility. As a paradigm for understanding cancer, tumoral evolution is far better at explaining how cancer cells adapt. The early metastasis of cancer cells now provides all the conditions necessary for tumoral evolution to proceed: genetic diversity and selection pressure. There are millions of genetically distinct circulating tumor cells (CTCs) being subjected to a selection pressure exerted by the anti-cancer defenses. It's not the specific mutation that's important; it's the understanding of what *drives* these mutations. Why are these genes mutating? Because their very survival depends upon it.

Even by the time of earliest diagnosis, the primary tumor has been shedding millions of CTCs daily, and micrometastases may have already been established.[12] CTCs can break off as single cells or as clusters that may work together to enhance mutual survival.[13]

While most CTCs are instantly killed as they leave the primary tumor site, very occasionally a rare genetic mutant survives its harrowing ordeal in the bloodstream—just barely. But these cancer cells do not yet have the ability to survive on the hostile shores of the liver, bone, or lung. When a cancer cell lands on a distant shore, it is immediately slaughtered. Other CTCs continue to circulate through the body, searching desperately for a safe harbor. At last, a few survive long enough to find an oasis, their site of origin: the primary tumor.

These prodigal sons of cancer return to settle back in their ancestral home, which provides them a haven. The returning CTCs require no new genetic adaptations to survive and thrive in the tumor microenvironment. In this tumor sanctuary, the acidic, hypoxic environment suppresses the immune surveillance that had been obliterating them out in the bloodstream.[14] This phenomenon is referred to as tumor self-seeding. Cell line models for breast, colon, and melanoma cancers have already proven its existence.[15] When tumors reseed their own primary site, they are ensconced in a protective environment—a nursery for returning cancer cells.

However, when these cancer progeny return, they are not the same wide-eyed innocents that originally left the primary tumor. Their companions have been killed. Only those that could activate their inner, most ruthless survivalist, the unicellular organism within, could return. Imagine a shoplifter sentenced to five years in the gulags of Siberia. He may have been a nice guy to start, but the horrors of the labor camp turned him into a hardened criminal. He's evolved. It's the same case with the cancer cell that returns to reseed the primary tumor. Only the cancer cells tough enough and adaptable enough to survive return. The conditions are ripe for natural selection. Cancer cells are genetically diverse and exposed to a selection pressure: being able to survive in the bloodstream.

The returning CTCs reinfiltrate the established tumor and outcompete the less aggressive original parental cancer cells. The primary tumor becomes replaced with a more aggressive strain, and the new CTCs being shed into the bloodstream are now the progeny of this strain. While the journey through the bloodstream is still treacherous, these more aggressive cells can now withstand it just a tiny bit better than the previous generation.

This is not the end of the story; it's only the beginning. The evolution of the cancer cell is an iterative process (see Figure 19.1).

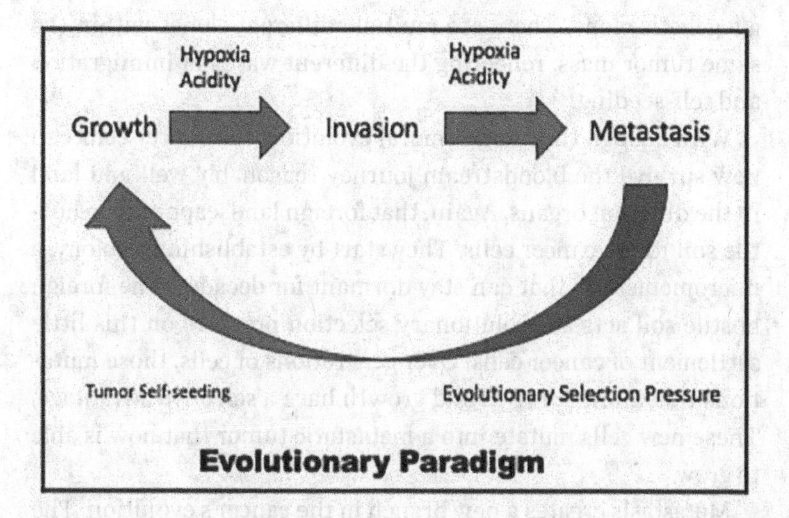

Figure 19.1

The new CTCs face another round of intense selection pressure. Most of the new CTCs are also killed, but once again, a few rare mutants who developed enhanced survival mechanisms in the bloodstream return to reseed the primary tumor site. This cycle repeats over and over again, usually over years or decades, each cycle bringing about new mutations that allow better survival in the bloodstream. The primary tumor is recolonized by its own metastatic progeny. These genetic mutations are anything but random—this is the Darwinian process of natural selection. The cancer cells are evolving.

Each iteration selects for more survivability and aggression. Over time, the cancer selects the hundreds of mutations necessary to successfully metastasize through the blood and set up a new colony of cells. Tumor self-seeding amplifies the most aggressive clones and allows selective enhancement of survivability traits.

The self-seeding process explains the genetic diversity with the primary tumor site—known as intratumoral heterogeneity (ITH). The cancer cells within a single tumor are not made up of a single genetic clone. There are multiple different clones within the same tumor mass, reflecting the different waves of immigration and self-seeding.

With enough time and tumoral evolution, the cancer cells can now survive the bloodstream journey reasonably well and land in the different organs. Again, that foreign landscape will be hostile soil to the cancer cells. They start by establishing a colony, a micrometastasis that can stay dormant for decades. The foreign hostile soil acts as evolutionary selection pressure on this little settlement of cancer cells. Over generations of cells, those mutations that allow survival and growth have a selective advantage. These new cells mutate into a metastatic tumor that now is able to grow.

Metastasis creates a new branch in the cancer's evolution. The metastatic site shows completely different genetics from the primary tumor,[16] which reflects the particular challenges of surviving in that metastatic site. Because most cancer cells die when trying to metastasize, very few cells survive, and this restricts genetic diversity,[17] a phenomenon known in biology as "bottlenecking."

At last, we have a working paradigm of how cancer develops, from its inception right through to metastasis. The factor that shapes cancer from beginning to end is the most powerful force in biology: evolution. It is not a random patchwork of different mutations. At all stages, the driving force behind cancer is the most primal directive of all life on earth: survival of the cell.

Cancer paradigm 3.0, the evolutionary model, can be divided into three phases:

1. Transformation: The normal cell's first step toward cancer is an evolutionary response to chronic, sublethal injury. The cancer phenotype develops as a survivalist mechanism that requires rejecting multicellular life. This is the seed of cancer.
2. Progression: The nutrient sensors insulin, mTOR, and AMPK influence growth factor availability and provide a fertile environment for cancerous proliferation. This is the soil.
3. Metastasis: The early shedding of cancer cells into the bloodstream exposes the cells to intense selection pressure for survival. As the primary tumor site is reseeded by its own progeny, natural selection pressure favors more aggressive and primitive traits.

20

THE STRANGE STORY OF CANCER

OUR UNDERSTANDING OF the fascinating and strange story of cancer has undergone several major revisions. The evolutionary paradigm offers astonishing new insight into cancer's genesis, from transformation, progression, and metastasis to treatment and recurrence. The focus shifted from purely genetics (the seed) to the environment (the soil) and the interaction of the two. In evolutionary biology, the environment plays the most important role in determining which mutations are beneficial and which are harmful. We've covered a lot of new ground, so before moving onto therapeutic implications, let's briefly recap the story of cancer, using the example of a smoker who develops a lung cancer that metastasizes to the liver.

Several billion years ago, the earliest organisms were simple single-cell prokaryotes. As they evolved into the more complex eukaryotes, increased cellular cooperation led to the development of multicellular organisms. These larger and more complex animals dominated against the simpler, smaller organisms, just as cities dominate against isolated individuals. But this required a fundamental and monumental shift in cellular priorities.

The previous genetic programming honed cells to compete against one another in order to survive. Now cells needed coordination and cooperation. Your liver does not try to kill your lung; the two help each other. Other cells were friends, not food. By cooperating, the entire organism gains a huge advantage of specialization. A single cell could not possibly learn how to read Shakespeare.

As cells transitioned from rivals to teammates, new rules were needed. The ancestral competitive playbook was not erased. Instead, new programs (genes) were added on top of the old ones to change and control them. Tumor suppressor genes suppressed the old "grow at all costs" programs. Oncogenes produced growth factors that activated old growth programs, but only at the correct time and place. Cancer appears precisely at this junction between unicellularity and multicellularity, or cellular competition versus cooperation.

Think of it as akin to training a bear to dance. A wild bear can be persuaded by early extensive training to do stupid human tricks, such as dance and wear a tutu. The original "wild animal" programming in the bear lies dormant but intact. A new "dance and wear a tutu" program has simply been overlaid. When the bear is provoked, it stops dancing and reverts to behaving as a wild animal, albeit with the tutu still on. Cancer is that inner wild beast, a ferocious competitor and survivor.

Unicellular genes that enhance competition and survival are *precisely those genes that cause cancer in multicellular organisms.* The seed of cancer already exists in every multicellular organism, because it is simply a remnant of our evolutionary past. When the new rules break down, the old unicellular behaviors reassert themselves. The seed of cancer grows, is immortal, moves around, and uses the Warburg effect. This is an ancient tool kit of survival responses. These are the hallmarks of cancer. This is the new invasive species known as cancer.

Because cancer is a dormant part of ourselves, and a known ever-present danger, multicell organisms have evolved potent anticancer strategies, including DNA repair, apoptosis, the Hayflick limit, and immune surveillance. Natural killer (NK) cells are a part of our innate immune system that—surprise, surprise—naturally kills tumor cells. NK cells constantly patrol the body like a vigilant police force, looking for potentially cancerous cells. Orders? Shoot to kill. If the anticancer defenses are impaired, cancer may flourish.

Multicellular organisms suppress competition between their own cells in the same way that a society enforces rules for individuals to cooperate, not compete. We line up at the movies instead of shoving each other out of the way. Cancer results from the loss of cooperation between cells and proceeds in three stages: transformation, progression, and metastasis.

TRANSFORMATION

Why would a cell reject polite multicellular society to compete on its own? Because its own survival is at stake. When the normal law and order of the multicellular society fails, cancer cells evolve to survive against chronic sublethal damage. If the injury is too significant, cells simply die. If the injury is too slight, the damage is simply repaired. With just the right amount of chronic injury, cells become as frantic as a mouse in a trap, forced to find a way to survive.

In our example, the constant damage from cigarette smoke forces lung cells to face an existential crisis. Some cells are killed outright. Some cells are untouched. But there's a substantial proportion of cells that endure chronic sublethal damage, and these activate wound healing. The normal rules of multicellular society have broken down. It soon becomes every cell for itself.

So, the damaged lung cell faces a dilemma: should it continue cooperating with other cells, its normal modus operandi? Without the normal law and order, this will likely lead to its own death. The alternative is to compete for its own survival and ignore the normal rules of multicellular society. Caught between the devil and the deep blue sea, some cells revert to competition.

The chronic sublethal smoke damage exerts a potent selection pressure, and the cell must activate the rarely used ancient survival subroutines from unicellularity to save itself. Natural selection favors certain genes for survival. Growth, immortality, movement, glycolysis (the Warburg effect)—those cells that don't adapt and dust off the old survival playbook don't survive. The cancerous transformation has just been made.

Mutations are not randomly accumulated, but carefully selected by a Darwinian evolutionary process. Mutations that activate the old unicellular, survivalist kernel of programming improve the odds of cell survival against this chronic smoke damage. As they turn down the backward path toward unicellularity, normal lung cells transform into cancerous cells.

The tumor, a little cluster of cancer cells, starts to grow in the lung. These radical cells represent danger to the organism, so the body activates its highly evolved and potent anticancer mechanisms to maintain multicellular law and order. Most of the time, the body kills this invasive species, wiping it out before a beachhead is established. In these cases, the cancer is killed before it can even be detected.

But the person is still smoking, and the chronic cell damage persists. Occasionally, a rare mutation allows cancer to survive the body's anticancer defenses. It doesn't flourish, but it's not totally dead. The little cancer develops genetic variation (intratumoral heterogeneity), which permits branched-chain evolution and natural selection. Over time, tumoral evolution slowly selects for the expression of more survival traits, which can take decades. Cancer uncovers previously suppressed abilities through genetic mutation, but again, these *atavisms are not random*. Rather, selective evolutionary pressure made possible by ITH is the guiding force for these mutations.

PROGRESSION

As the tumor grows, it faces new challenges that require new solutions that cannot be found in the old survivalist playbook. Parts of the growing tumor are too far from their vital supply lines, the oxygen-carrying blood vessels. This activates hypoxia-inducible factor 1 (HIF1) which stimulates the growth of new blood vessels. This is not part of the original survival subroutine, but evolved to support tumor growth.

In addition, HIF1 encourages cell movement and the Warburg effect. The cancer cell produces lactic acid, which is dumped into the surrounding microenvironment. The acid not only suppresses normal immune cell function, but also degrades the supporting structures, making it easier for cancer cells to invade through the basement membrane, enter local tissues, and eventually disseminate through the bloodstream. The damage from the acidity attracts inflammatory cells that produce growth factors. Normal cells fare poorly in this hypoxic, acidic environment, whereas cancer cells do relatively well. In the land of the blind, the one-eyed man is king.

Cancer seeds are everywhere, but they are irrelevant without fertile soil. Multicell organisms tightly control growth, just as cities tightly control growth through building permits. But certain conditions, such as the ready availability of nutrients and especially glucose, allow easy growth of both normal and cancer cells.

The human body uses three main nutrient-sensing pathways, insulin, mTOR, and AMPK, which also act as growth factors. When the body senses more available nutrients (high insulin, mTOR, and low AMPK), conditions favor growth, which favors cancer cells. The lung cancer is now not just surviving, but also finding good soil to grow. But it's getting a little too big and unwieldy—time to leave home.

METASTASIS

Cells are normally stuck to their place of origin by surface adhesion molecules. But cancer cells that can't move can't grow. For an invasive species, movement is a perfectly normal behavior. Cancer cells break off the primary tumor to find more room to grow. This happens early in cancer's course, as the circulating tumor cells (CTCs) consume nutrients rapidly and are very quickly driven by the increased competition for resources. This new environmental stress creates new evolutionary selection pressures.

Unfortunately, these CTCs find out that the bloodstream is a terribly hostile environment, and most just die. But not all. One day, a rare genetic mutant arises that is able to survive both the immune cell attack and the travel through the bloodstream long enough to circulate back to the original tumor site in the lung.

As it returns home, it finds a sanctuary against all the scary things trying to kill it, and it recovers. The tumor has just self-seeded. But this returning cancer is more aggressive and just a tiny bit better able to survive in the bloodstream. This more aggressive variant multiplies within the safety of the primary tumor site. It dominates and outcompetes the incumbent cancer cells. Driven by the incessant hunger to grow, the new lung cancer strain starts shedding circulating tumor cells into the blood. This cycle of tumor self-seeding and metastasis repeats over and over, with the cancer continually evolving over time its ability to survive the bloodstream.

But cancer cells can't circulate forever in the blood like a raft adrift in the ocean. After barely surviving their harrowing trip through the bloodstream, the cells land on the hostile shore of some other tissue. These lung cancer cells may land on the kidney, liver, bones, or the brain. Almost all these cancer cells are immediately wiped out in this new hostile, alien environment. A lung cell can't survive in the kidney or the liver or the bones or the brain. It's a fish out of water.

Millions of lung cancer cells die on these alien shores. But the primary tumor is still shedding millions more cancer cells into the blood just behind them. Wave after wave of cancer cells die. It takes more than a few tweaks to the original lung cancer cell to adapt to the entirely new environment of the liver.

Eventually, a rare genetic mutation allows the lung cancer cells to reach the new distant shore of other organs, like the liver, and manage to survive. They may not thrive, but at least they're not dead. This micrometastasis is so small that it is undetectable and may lie dormant for decades. Invasion and metastasis are difficult skills to master, and most cancers fail.

Given enough time, Darwinian evolutionary processes select a rare genetic variant to flourish, and the little outpost of metastatic cancer cells grows. The cancer has just become metastatic, and the prognosis for the patient has just plummeted from good to decent. This slow process from initial carcinogenesis to metastasis takes decades.

Once the cancer is discovered, the patient begins a regimen of chemotherapy, radiation, hormones, or surgery. These weapons of cellular mass destruction annihilate the embattled cancerous foe, but their efficacy is limited by their damage to normal cells, which causes side effects. The cancer may appear to have gone into remission, but if even a few cancer cells survive, the treatment puts a new selection pressure on the cancer. Eventually, a rare variation arises that shows resistance to the treatments given. Those resistant cells replicate and flourish. The cancer has recurred. And now it's resistant to treatment. The prognosis has now just plummeted from decent to poor.

CANCER PARADIGMS

The evolutionary/ecological model of cancer, with tumoral evolution and self-seeding, reflects the complex, dynamic, and ever-evolving ecosystem of cancer. It considers not only the cancer cell itself but also its relationship with other cells and with its environment. Population dynamics, evolution, and selection pressure are key elements of this new model of cancer. The evolutionary model is a mosaic, not a monochrome like the SMT.

Our understanding of cancer has moved through three major paradigms, each of which has produced revolutionary treatments and improved understanding, but that journey has also revealed our knowledge gaps with regard to this ancient enemy. Cancer paradigm 1.0 understood cancer purely as a disease of excessive, uncontrolled growth. The logical solution is to kill those areas of excessive growth with poison (chemotherapy), burning (radiation), and/or cutting (surgery). But this paradigm reached its limits by the mid-1970s.

Cancer does one thing better than anything else: grow. The treatments of cancer paradigm 1.0 were trying to beat cancer at its own game. We were attacking cancer's strength, not its weakness. If cancer were a crab, we were attacking it head-on, trying to crack its tough shell, while suffering from its vicious claws. Cancer paradigm 1.0 ultimately failed to explain *why* these cells were growing excessively.

Cancer paradigm 2.0 explained that accumulated genetic mutations caused excessive growth. A few mutations in a few key growth genes accounted for cancerous growth and worked well for some cancers, but problems soon emerged. As we chased down the various genetic mutations, there were more than just the expected few. There were *millions*.

Cancers are master genetic shape-shifters, mutating frequently and better than anything else in the known universe. So, targeting a specific mutation to kill cancer is largely futile, because the cancer can simply mutate again. Cancer is a dynamic disease, so any single static treatment, whether chemotherapy or a genetically targeted drug, often fails. Instead, treatments may introduce a new selection pressure; just as constant antibiotic use can build up bacterial resistance. Cancer is the ultimate survivor, having evolved over billions of years to evade threats. Once again, we are attacking cancer's strength and hoping for the best. With an almost infinite number of combinations of mutations, genetically targeted precision treatment was revealed to be a pipe dream.

As with its predecessor, cancer paradigm 2.0 (the genetic model) failed because it neglected to explain *why*. Why were cells mutating? With no answer, this cancer paradigm reached its limits by the early 2000s. We had zoomed in so closely to the specific genetics of cancer that we missed the importance of the environments and interactions between cells. We had missed the forest for the trees.

A NEW DAWN

This led us finally to the current evolutionary/ecological theory of cancer: cancer paradigm 3.0. Darwinian evolution is the only known force in the biological universe that can create and coordinate the sheer number of mutations needed for cancer. The quest for cellular survival drives the accumulation of tens or hundreds of mutations seen in each cancer.

Cancer is not only a seed problem but also a soil problem. Ecology is the study of the relationship between one organism and another, as well as these organisms' relationship to their environment. The evolutionary theory does not negate the importance of the genetic paradigm, but it expands it to include both the seed and the soil. Cancer is not just a genetic disease. The evolutionary/ecologic paradigm recognizes the importance of cell-to-cell interactions and interactions with the environment, making it a far more dynamic, inclusive, and comprehensive theory of cancer. Evolutionary biology links carcinogenesis, progression, and metastasis, whereas genetics considers them all as separate issues.

This idea is not new; it just needed to be rediscovered. "Cancer is no more a disease of cells than a traffic jam is a disease of cars," wrote cancer researcher D. W. Smithers in 1962. A traffic jam results from the interaction between the car, neighboring cars, and the environment. If you look only at each individual car—Are the brakes working? Was it recently serviced?—you will fail to find the problem.

Similarly, cancer is not only a genetic disease, but also an ecological disease. The environment plays a huge role in determining whether a cancer grows. Under certain conditions, such as high insulin levels, cancer will thrive, while under other conditions, it will fail to establish itself.

This new understanding of cancer has major implications for cancer prevention and treatment. An entire new front on the war on cancer has now been opened. We've been able to identify opportunistic targets to break the stalemate of the past fifty years. A firestorm of research has enabled the development of entirely new weapons to defeat cancer. Life expectancies are rising. Cancer death rates are falling. For the first time in living memory, cancer may be on the retreat.

PART VI

TREATMENT IMPLICATIONS

21

CANCER PREVENTION AND SCREENING

TODAY, HEART DISEASE is the leading cause of death in the United States, with cancer as the perpetual runner-up in this morbid race. But that may soon change. From 1969 to 2014, death from heart disease declined by 68.4 percent among men and 67.6 percent in women, due to both improved prevention and improved treatment.[1]

Yet, over that same period, cancer deaths declined by a relatively minimal 21.9 percent for men and 15.6 percent for women—less than one third the rate at which heart disease declined (see Figures 21.1 and 21.2). In 1969, the risk of dying from heart disease was two to three times higher than the risk of dying from cancer. In 2019, those risks were virtually equal.[2] As recently as 2000, cancer was the leading cause of death in only two states (Alaska and Minnesota). In 2014, fully twenty-two states reported that cancer was the leading cause of death.[3] According to the American Cancer Society, the lifetime risk of being diagnosed with cancer is more than one in three,[4] and worse, many of the obesity-related cancers are rising in prevalence and considered avoidable.

While progress against cancer lags far behind that for heart disease, there are still some bright spots. Total cancer deaths peaked in 1991 and fell steadily by 27 percent in the twenty-five years to 2016, due mostly to the reduction in lung cancer deaths thanks to the increased regulation of the tobacco industry and effective antismoking campaigns.

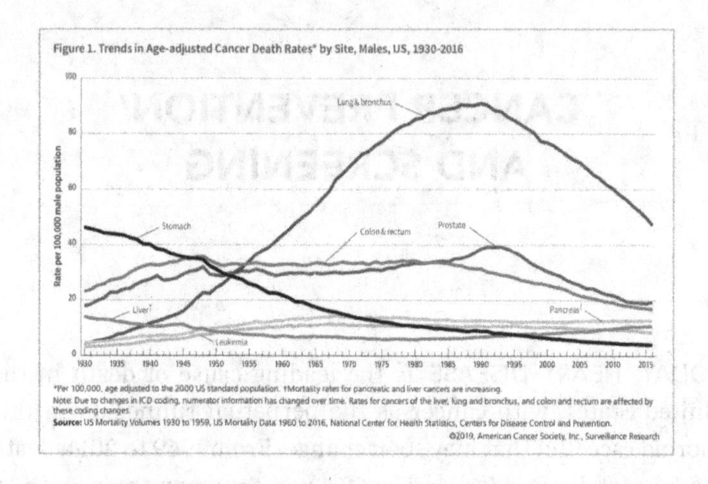

Figure 21.1

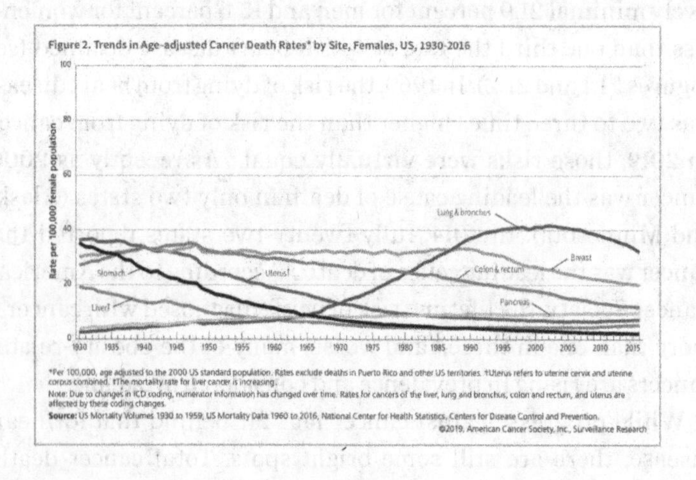

Figure 21.2

Prevention is the surest route to defeating cancer and is responsible for this significant victory in the reduction in lung cancer deaths. The popularity of smoking tobacco in the United States began to increase around 1900, gaining momentum during World Wars I and II, and peaked in 1964, when 42 percent of Americans smoked.[5] Smoking was fashionable, common, and seemingly harmless. Many physicians smoked, along with much of the rest of the male population. Smoking in women was decidedly less popular in the 1960s, but it gained steam afterward, ironically as a symbol of female empowerment.

The seminal event in the history of cancer prevention was the 1964 declaration by Luther Terry, then the surgeon general of the U.S. Public Health Service, that smoking caused lung cancer. Cigarette smoking is estimated to account for 81 percent of lung cancer (see Figure 21.3). Terry, himself a long-term smoker, almost singlehandedly saved hundreds of millions of lives.

Public perception of smoking changed gradually but irrevocably after the 1964 report. Just a year later, new legislation mandated that warning labels be placed on cigarette packaging, to let consumers know that smoking was hazardous to their health. Other public health measures included restricting cigarette advertising, especially to young people. Smoking rates in the United States steadily declined over the ensuing years, reaching 15.5 percent by 2016.[6]

Lung cancer deaths followed an identical trajectory, delayed by the twenty-five years or so that it takes to develop cancer. From 1990 to 2016, lung cancer deaths fell 48 percent in men. Smoking in women started later than in men, and therefore showed a slower uptake and decline.

But cigarette smoking contributes to more than just lung cancer. The chronic irritation caused by tobacco smoke causes at least twelve other cancers and increases the risk of heart disease, stroke, and chronic lung disease.[7]

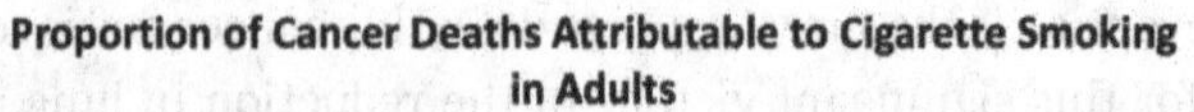

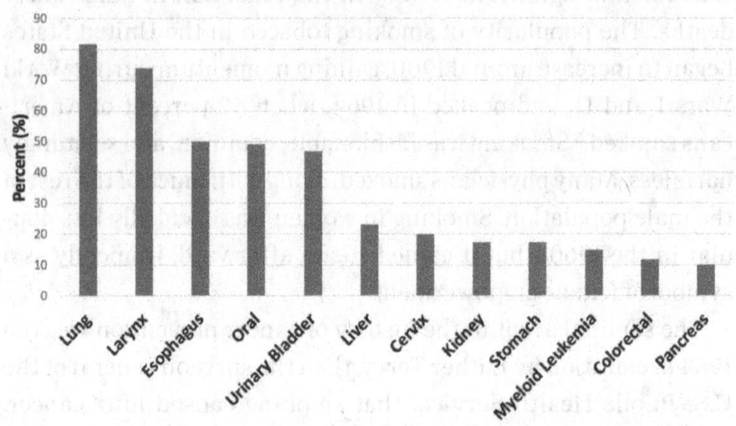

Data from: Islami F et al. CA Cancer J Clin, 2018;68:31-54

Figure 21.3

Lung cancer was still responsible for most cancer deaths in 2019, for both men and women, although the numbers are dropping significantly. There is little doubt about the best way to treat lung cancer: stop smoking. Nothing else comes anywhere close. It's a matter of execution, not knowledge.

Cancers caused by infectious agents like bacteria and viruses have also been declining steadily. Stomach cancer has fallen steadily since 1930, as improving public sanitation dramatically decreased the prevalence of *H. pylori*. Overcrowding and poor sanitation persisted in Asia for much of the first half of the twentieth century, leading to endemic *H. pylori* infection and high rates of stomach cancer. In Japan, for example, death from stomach cancer did not start to fall until the mid-1960s.[8]

Liver cancer fell steadily in America from 1930 to 1980, due to major advances in the identification and prevention of the viruses hepatitis B and C. Both viruses are less common in America compared to Asia, where hepatitis B has stayed endemic due to maternal transmission. Deaths from liver cancer in Japan did not peak until the 1990s.[9] Today, widespread vaccination for hepatitis B and effective antiviral medications for hepatitis C have led to the hope that liver cancer may follow its persistent downtrend.

The recent news on liver cancer is not good, but for a completely different reason. Over the last forty years in the United States, liver cancer diagnoses have tripled and deaths have more than doubled. There is little mystery as to why this is so: liver cancer is one of the obesity-related cancers, with obese and overweight individuals suffering almost double the risk compared to healthy-weight individuals.[10] Fatty liver disease can cause chronic inflammation, leading to cirrhosis and liver cancer. Pancreatic cancer, also an obesity-related cancer, is increasing in incidence, by about 1 percent per year from 2006 to 2015.

SCREENING

After lung cancer, the next biggest cancer killers are prostate cancer (men) or breast cancer (women) and then colorectal cancer in third place. In all three types of cancer, no single major causative agent has yet been identified, although breast and colon cancer are the two most prominent obesity-related cancers. Without knowing causation, these cancers cannot be entirely prevented. The next best step may be early detection and treatment through screening. When they are caught early, the five-year survival rate for breast, colorectal, and prostate cancer exceeds 90 percent (see Figure 21.4).

Once the cancers become metastatic, however, these survival rates drop to less than 30 percent. The key, then, is to reduce the number of patients who present with late-stage disease. One well-tested strategy is to screen for early disease with the hope that we can reduce the number of people presenting with the more lethal,

FIVE-YEAR SURVIVAL (PERCENT), UNITED STATES, 2008–2014

Metastatic Spread	Local	Regional	Distant
Breast (Female)	99	85	27
Colorectal	90	71	14
Prostate	> 99	> 99	30
Esophagus	45	24	5
Pancreas	34	12	3

American Cancer Society, Facts & Figures, 2019.

Figure 21.4

late-stage disease. While several large population-based screening programs have been wildly successful in reducing the burden of cervical and colorectal cancer, screening for three other common cancers (breast, prostate, and thyroid) have been decidedly less successful. Unfortunately, simply catching more early-stage disease is not useful by itself. In the first two cases (breast and colorectal cancer), screening has reduced late-stage disease, but in the latter three cases (prostate, esophageal, and pancreatic cancer) it has not, and that makes all the difference in the world.

CERVICAL CANCER

Deaths from cervical cancer have dropped dramatically since the 1940s, largely due to the introduction of the Pap smear. It's now known that 70 percent of cervical cancer is attributable to two strains (16 and 18) of the human papillomavirus (HPV), a disease transmitted largely through sexual contact. After infection by these cancer-causing subtypes of HPV, the cervix sheds abnormal cells for several years prior to the development of invasive cervical cancer.[11]

In 1928, gynecologist Dr. George N. Papanicolaou discovered that when the cervix was scraped with a small brush and the sample cells obtained were examined under a microscope, previously hidden cancer cells could be detected.[12] Women had no symptoms at this stage, and felt entirely well. By 1939, Papanicolaou was routinely sampling cells from all women admitted to the obstetrical and gynecological service at his New York hospital. In 1941, his landmark research paper described how the "Pap" smear could detect unsuspected precancerous lesions.[13] With the cancer diagnosed early, removal of these lesions halted further progression toward a full-blown cancer. Today, cervical cancer serves as the poster child of screening programs, being one of the earliest and most successful cancer interventions to date.

Throughout the 1940s and '50s, the American Cancer Society enthusiastically promoted the Pap smear for mass screening. It trained physicians and pathologists in its use and established a series of cancer detection clinics, focused mainly on cervical cancer.[14] Deaths from cervical cancer dropped an estimated 71 percent from 1969 to 2016. The good news is that large-scale, population-based vaccination programs have been developed and approved in many countries. Preventing viral transmission of HPV 16 and 18 promises to lower the rate of cervical cancer even further.

COLORECTAL CANCER

Population-based screening for colorectal cancer is another great example of how to increase the odds of successful cancer treatment, as deaths from colorectal cancer have trended steadily downward since the mid-1980s, due largely to population-wide screening programs.

In 1927, researchers discovered that normal colon tissue did not transition directly into colorectal cancer, but went through an initial precancerous phase called an adenomatous polyp.[15] The polyp stage lasted for years or even decades before transforming into an invasive cancer. Before the 1960s, detecting adenomatous polyps was no easy task.

Digital rectal exams were historically the preferred method of screening. As part of the routine physical examination, patients heard the dreaded snapping of the latex glove and the stern instructions to bend over, just before the physician's finger was inserted into their rectum, to feel around for anything unusual. Not only was this exam unpopular, it was also virtually useless, as the human colon is approximately five feet long, and the physician could feel only the first few inches. Luckily, advances in technology would eventually provide a better solution.

In the 1940s, a rigid camera called a sigmoidoscope was developed. It was inserted into the sigmoid colon, the last fifteen inches or so of the large intestine. It was a difficult and painful procedure, but nevertheless, starting in 1948, large-scale screening programs using this primitive technology resulted in a jaw-dropping 85 percent reduced rate of colorectal cancer over twenty-five years.[16] The concept was proven, but the screening procedure was not generally acceptable to the public.

Improved medical technology changed the game once more. Polyps may bleed intermittently, allowing small amounts of blood into the stool that are not immediately visible to the naked eye. In the late 1960s, fecal occult blood testing (FOBT) was developed, allowing otherwise invisible amounts of blood to be detected and serving as an early warning sign of colorectal cancer. This required only a simple stool sample, and no tubes needed to be inserted into any orifices.

By the middle of the 1970s, flexible colonoscopy was developed. Instead of a rigid scope with limited reach, the colonoscope was flexible, allowing easier passage throughout the entire colon, with its many twists and turns. A positive test for fecal occult blood could now be followed up by a relatively simple colonoscopy that could both detect and remove polyps.[17]

In 1993, findings from the National Polyp Study proved that this combined approach reduced colorectal cancer by a stunning 76 to 90 percent[18] and cancer deaths by 51 percent.[19] The 1993 Minnesota Colon Cancer Control Study confirmed that screening with fecal occult blood and colonoscopy reduced death from colorectal cancer by 33 percent.[20] This was not just a home run for cancer prevention and treatment; it was a grand slam.

Currently, the U.S. Preventive Services Task Force (USPSTF) recommends screening with FOBT or colonoscopy starting at age fifty and until age seventy-five. Colonoscopy has the advantage of being able to both detect the polyp and remove it immediately. Screening rates have increased from the 20 percent range in the 1990s to approximately 65 percent in the United States today.[21]

The increased screening has steadily decreased deaths from colorectal cancer, but the 2019 statistics contain some disturbing data: this cancer is increasing disproportionately in younger patients, likely related to the ongoing and increasing obesity crisis. The American Cancer Society estimates that 55 percent of colorectal cancers are attributable to modifiable risk factors, predominantly excess body weight. For those patients over age fifty-five, colorectal cancer decreased by 3.7 percent per year from 2006 to 2015; but for those younger than fifty-five, it increased by 1.8 percent.

Both cervical cancer and colorectal cancer progress in a fairly orderly fashion, from a premalignant condition to invasive cancer. This provides a window of opportunity where early detection and intervention can stop cancer's progression. Hopes were high that these successes could be repeated for the other big killers, breast and prostate cancer, through mammography and prostate-specific antigen (PSA) blood testing, respectively.

BREAST CANCER

Breast cancer screening focuses on mammography, a type of X-ray, because breast self-examination is too variable and unreliable. For decades, cancer societies recommended annual screening mammography for women starting at age forty, and by all accounts, early screening seemed like a successful intervention. Breast cancer deaths peaked in 1989 and declined by 40 percent from 1989 to 2016. So, it is somewhat surprising that, recently, many countries are suggesting *less* screening, particularly for those ages forty to fifty.

In 2013, the Cochrane Library, acknowledged as a world expert in evidence-based medicine, reviewed all available data on mammography and concluded that it provided no overall benefit in preventing breast cancer deaths.[22] How could that be possible? But the Cochrane Library was not alone in its doubts.

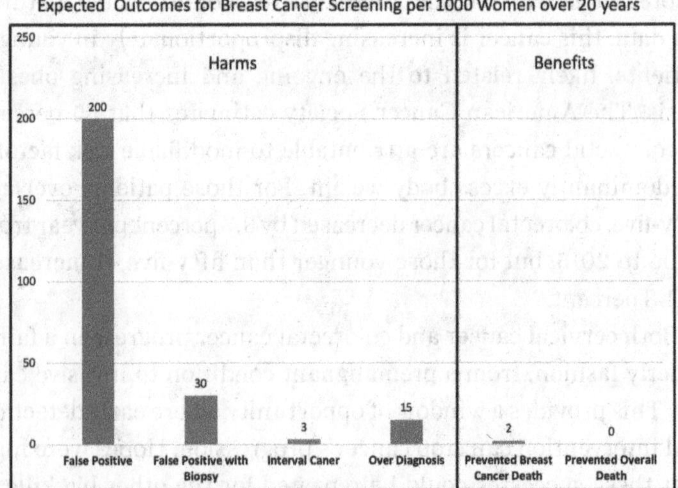

Figure 21.5

In 2014, the Swiss Medical Board noted how "non-obvious it was that the benefits outweighed the harms."[23] It was not immediately apparent to these experts that screening mammography did any good whatsoever. For a fifty-year-old woman, the Swiss Medical Board estimated that screening prevented only 1 breast cancer death for every 1,000 women screened. This means that the other 999 (or 99.9 percent) women did not benefit directly from mammography, but may have suffered from overdiagnosis.

By contrast, it was obvious to any casual observer that the Pap smear screening had dramatically reduced the burden of cervical cancer. No randomized trial was ever needed because the benefits were so clear. Where was the disconnect?

There are three main problems with breast cancer screening: lead time bias (see Figure 21.6), cancer deaths versus overall deaths, and failure to reduce late-stage disease. Much of the benefit of screening is illusory due a phenomenon known as lead time bias. Imagine two women who both develop breast cancer at age sixty and who both die of the disease at age seventy. The

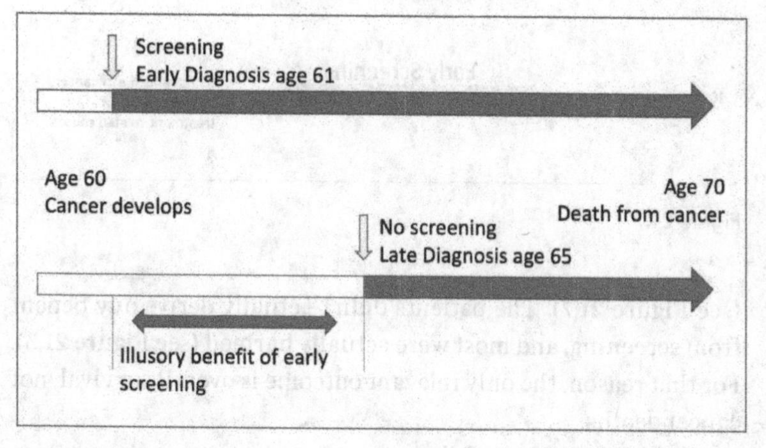

Figure 21.6

first undergoes screening and finds the disease at age sixty-one. The other does not undergo screening and finds the disease at age sixty-five. The first woman "survived" cancer for nine years, whereas the second "survived" cancer for only five years. Screening improved cancer survival by an apparent four years, but this is only an illusion.

The second problem is a methodologic issue. Many screening programs declare themselves a success by reducing cancer deaths rather than overall deaths. Why is this important? Suppose a group of one hundred people with cancer will die of the disease within the next five years, and early screening is completely ineffective. However, the toxicity of treatment (surgery, radiation, and chemotherapy) causes heart attacks and infections, killing twenty-five of those patients. Without screening, one hundred patients died of cancer, but with screening, only seventy-five died of cancer and twenty-five died of other causes. Screening reduced cancer deaths by 25 percent, but the benefits were entirely illusory

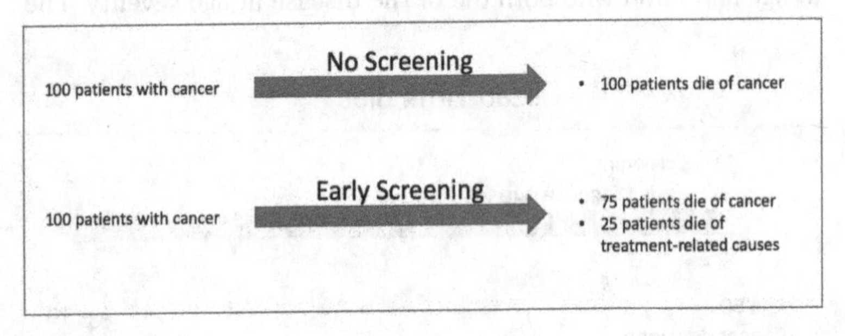

Figure 21.7

(see Figure 21.7). The patients didn't actually derive any benefit from screening, and most were actually harmed (see Figure 21.5). For that reason, the only relevant outcome is overall survival, not cancer deaths.

The third problem, failure to reduce late-stage diagnosis, is much more serious. Mammography detects plenty of early-stage breast cancer, exactly as intended. From 1976 to 2008, screening caught more than twice the number of early-stage breast cancers compared to the prior period before widespread screening (see Figure 21.8). Logically, you would expect that catching and treating the disease early should reduce the amount of breast cancer diagnosed at a late stage. *But it didn't*. The rate of late-stage cancer decreased by only a minuscule 8 percent.

Late-stage disease is highly lethal, whereas early-stage disease is highly curable. The early-stage cases being detected were unlikely to progress to more advanced disease. Only a tiny 6.6 percent of those early-stage cases were expected to progress to invasive cancer. In other words, in 93.4 percent of the early-stage cases detected by screening, treatment offered no discernible benefit. We were finding cases that didn't need to be treated. And we weren't significantly reducing the number of late-stage cases that are the deadliest.[24]

But why didn't the detection of more early-stage disease trans-

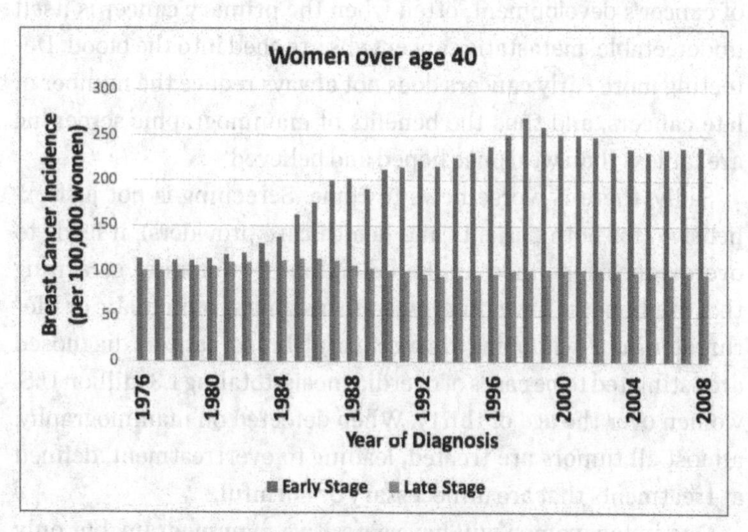

Figure 21.8

late into fewer overall late-stage cancers? In both cervical and colorectal cancer, progression follows an orderly and defined path, from microscopic tumor to larger tumor to metastasis. Therefore, virtually each small precancerous lesion detected and treated was one fewer late-stage cancer in the future. But breast cancer works differently.

The evolutionary model of cancer can help us understand why removing small cancers does not reduce the incidence of fully matured cancers. Cancer does not follow linear evolution, where one event follows the next in an orderly fashion. Instead, cancer follows branched-chain evolution. Thus, just as a tree may grow through a fence, blocking a single branch may not stop overall progression of the cancer.

Metastasis accounts for the majority of deaths from cancer. If metastasis occurs late, then early detection and treatment will lower the risk of metastasis. But the new paradigm of cancer recognizes that metastasis is an *early event.* Very early in the course of cancer's development, often when the primary cancer is itself undetectable, metastatic cancer cells are shed into the blood. Detecting more early cancers does not always reduce the number of late cancers, and thus the benefits of mammographic screening are far less than we'd once hoped and believed.

Sadly, there is worse news to come. Screening is not just expensive (for both patients and health care providers); it leads to overdiagnosis, defined as the detection of tumors at screening that might never have progressed to become symptomatic or life-threatening. A whopping *31 percent* of all breast cancers diagnosed are estimated to be cases of overdiagnosis, totaling 1.3 million U.S. women over the age of thirty. When detected on mammography, almost all tumors are treated, leading to overtreatment, defined as treatments that are unnecessary or harmful.

One in ten women will have a positive mammogram, but only 5 percent of those positives are actually found to be cancer. Put another way, 95 percent of women with a positive mammogram

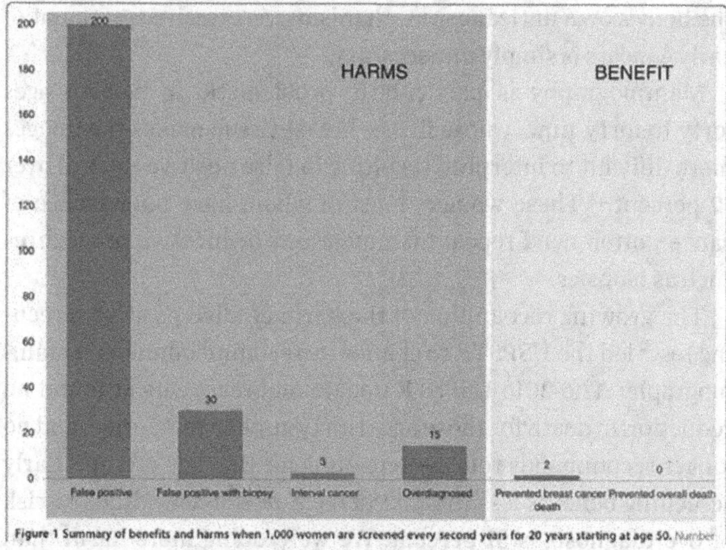

Figure 21.9

are subjected to invasive procedures for no eventual benefit. This includes biopsies, lumpectomies, and sometimes unnecessary chemotherapy. Women who get screening mammography are also treated more often with mastectomy and radiation. In the United States, the false positive rate is 30 to 50 percent.[25] In addition, it has been well established that women testing positive on mammography suffer psychological problems and poor quality of life, even up to three years after screening.

The majority of breast cancers diagnosed from mammogram screening are classified as ductal carcinoma in situ (DCIS), a very early stage of cancer. This diagnosis represents approximately 20 percent of all breast cancers, and its incidence has risen dramatically with the onset of widespread screening.[26] From 1983 to 2004, its incidence has risen tenfold. Many of these early-stage breast cancers are unlikely ever to progress to a dangerous state—as the evolutionary model explains, a cancer can be fully contained by the body's own anticancer mechanisms. Aggressive treatment of early cancers is simply unnecessary.

Mammography is particularly problematic in women ages forty to forty-nine, whose firmer breast tissue makes the images more difficult to interpret, resulting in false positive rates of over 12 percent.[27] These women, most of whom have no evidence of cancer, often need repeat mammograms or invasive procedures such as biopsies.

The growing recognition of the harm of false-positive screening tests led the USPSTF to change its recommendation on mammography. The 2016 USPSTF update on breast cancer found no reduction in death for those ages thirty-nine to forty-nine, and no longer recommends routine screening for that age group.[28] Early screening benefits less than 0.1 percent of women, while the risk of overdiagnosis is 31 percent. We were doing more harm than good (see Figure 21.9).

Mammography is not the holy grail of cancer screening we expected it to be. In men, a similar story has played out for prostate cancer.

PROSTATE CANCER

First identified in the 1960s, the protein called prostate-specific antigen (PSA) liquefies semen to allow sperm to swim freely. The PSA assay, which measures the amount of the antigen in the blood, was developed for law enforcement agencies in rape cases. By 1980, PSA was also found in the blood of patients with prostate cancer, raising the hope that PSA-based blood testing could become the male equivalent of the Pap smear.[29]

High PSA levels are not specific for prostate cancer, as they are also commonly found in men with enlarged or inflamed prostates. The FDA approved PSA-based screening for prostate cancer in 1986, setting the cutoff level at 4.0 ng/mL, as 85 percent of men had values lower than this. This meant that fifteen men out of one hundred had levels higher than 4.0 ng/mL and would undergo further prostate biopsy, of which between four and five would be diagnosed with aggressive prostate cancer.[30]

Enthusiasm for PSA-based screening exploded in the 1990s and 2000s. More than twenty million tests were performed each year, and more early prostate cancer was diagnosed than ever. In 1986, fewer than one third of men were diagnosed with early-stage disease confined to the prostate. By 2007, over two thirds of prostate cancer cases were early stage (see Figure 21.10). Death from prostate cancer began to drop, and it was looking like yet another feel-good cancer success story. But alas, the PSA story was not so simple.

Screening detected more early-stage prostate cancer, but did it improve overall survival? Three very large long-term studies of PSA-based screening answered that very question. In the United States, there was the Prostate, Lung, Colorectal, and Ovarian (PLCO) Cancer Screening Trial, which involved more than 76,000 men.[31] In Europe, there was the European Randomized Study of Screening for Prostate Cancer (ERSPC) study, with almost 182,000 study participants.[32] And in the United Kingdom, there was the PROTECT (Prostate Testing for Cancer and Treatment) study, with 408,825 participants.[33] With between 10 and 14.8 years of follow-up, none of these three enormous trials detected a discernible benefit with regard to overall survival to PSA-based testing. The USPSTF estimated the rate of overdiagnosis at 16.4 to 40.7 percent. The screening detected early, less aggressive prostate cancer, but did not reduce advanced disease. Once again, it seemed that screening was detecting the cancers that didn't require treatment.

Patients with a positive PSA are subjected to invasive procedures with significant side effects. Approximately 10 percent of men screened received at least one false-positive PSA screening result, leading to more than a million prostate biopsies being done

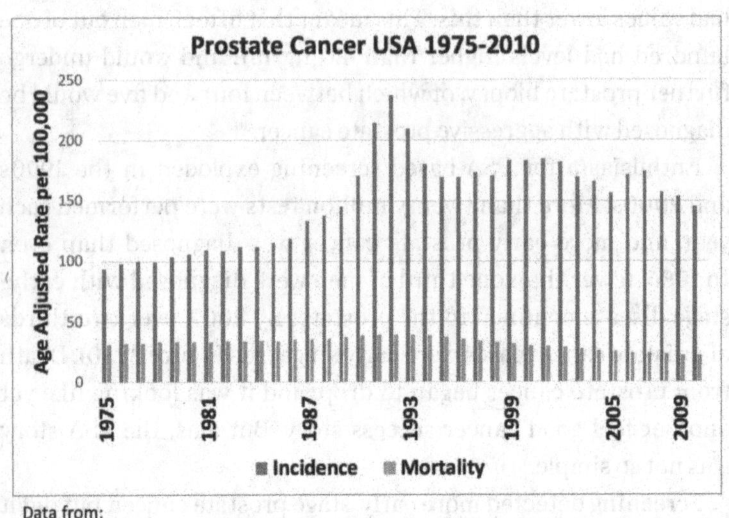

Data from:
https://seer.cancer.gov/archive/csr/1975_2010/results_merged/sect_23_prostate.pdf

Figure 21.10

per year.[34] Of the men in the PLCO screening group, 12.6 percent underwent one or more biopsies. Of those, 2 to 5 percent experienced biopsy-related complications. Given the huge numbers of men undergoing screening, this is a massive number of complications. As with mammography, overdiagnosis is a major problem. Men diagnosed with prostate cancer are more likely to have a heart attack or commit suicide in the year after diagnosis.[35]

In 2012, the USPSTF recommended against PSA-based screening, noting with moderate certainty that the benefits do not outweigh the harms. In 2018, the USPSTF again reviewed the evidence and recommended against routine screening.[36] The task force felt moderately certain that screening with PSA-based methods is worse than doing nothing at all.

According to USPSTF guidelines, men younger than fifty-five or older than seventy should not be tested for PSA. Between ages fifty-five and sixty-nine, it is an individual choice. They note that "Screening offers a small potential benefit of reducing the chance of death from prostate cancer in some men. However, many men will experience potential harms of screening."[37] Not exactly a screaming endorsement.

THYROID CANCER

In 1999, South Korea embraced national screening in an effort in its own "war on cancer." Screening for breast, cervical, colorectal, gastric, and hepatic cancer was provided free of charge to the entire population. Screening for thyroid cancer was not included, but for a small surcharge, approximately thirty to fifty dollars, patients could elect to have a neck ultrasound performed as well. By 2011, thyroid cancer was being diagnosed fifteen times more frequently than in 1993 (see Figure 21.11).[38] In almost all diagnosed cases, the thyroid was either partially or completely removed. This treatment was not without consequences. There was an

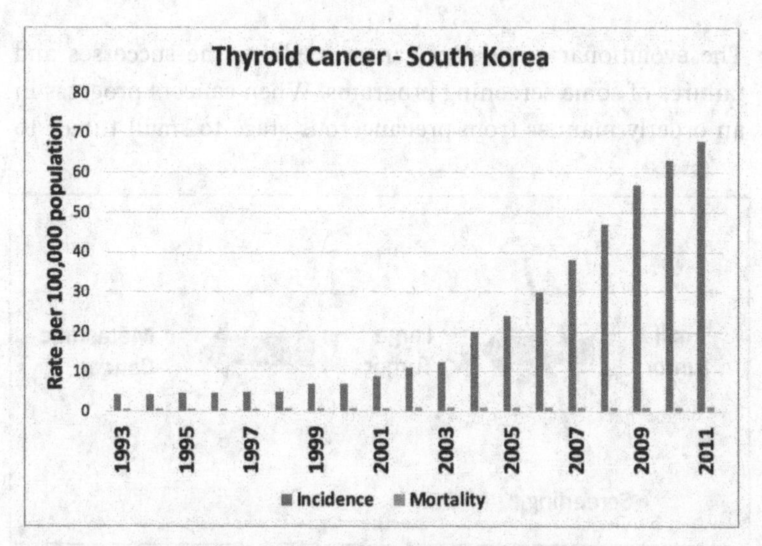

Data from: N Engl J Med 2014;371:1765-1767

Figure 21.11

11 percent risk of reduced parathyroid function and a 2 percent risk of paralysis of the vocal cords due to nerve damage.

Despite this intensive effort to eradicate early-stage thyroid cancer, the risk of death from thyroid cancer was virtually unchanged. Simply put, it was a classic case of overdiagnosis—most of the thyroid cancers detected in screening did not require treatment. Finding and treating early disease is not useful. Only reducing the incidence of late-stage disease is useful, and these are not necessarily the same thing due to the occurrence of early metastasis. By some estimates, as many as one third of all adults have evidence of thyroid cancer, but the vast majority of these cancers do not produce symptoms or cause health problems.[39] Finding and treating cancers that don't need to be treated is not a useful strategy.

CONCLUSIONS

The evolutionary model of cancer explains the successes and failures of some screening programs. When cancers progress in an orderly manner from precancerous stage to small tumor to

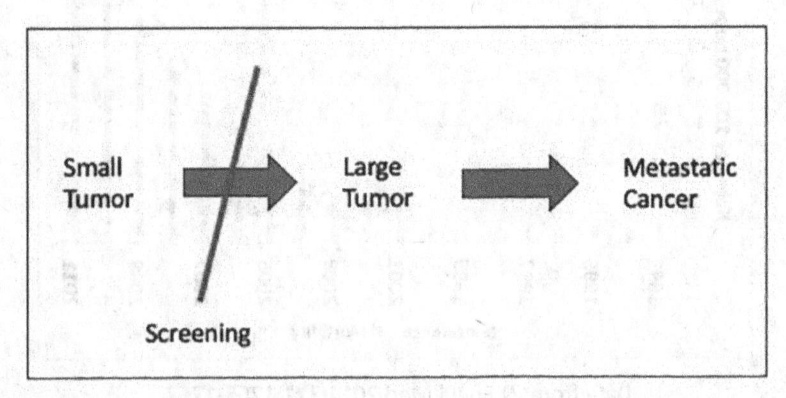

Figure 21.12

large tumor to metastasis, screening succeeds (see Figure 21.12). Screening to remove early cancers prevents the development of late cancers, and this saves lives.

But if removal of early-stage cancer does not reduce late-stage cancer, screening is unsuccessful, and overdiagnosis becomes a problem. Not every early-stage cancer needs to be destroyed, as many small cancers are contained by the immune system and will never pose a serious health threat. With toxic treatments such as surgery, radiation, and chemotherapy, the treatment may be worse than the disease.

Consider the normal bacteria in your gut, known as the microbiome. Must every bacterium in your body be eradicated? No. Most bacteria living in your gastrointestinal tract are neutral or even good. Probiotic supplements and foods containing live cultures (e.g., yogurt) derive much of their proposed benefits by encouraging the growth of these "good bacteria." Killing every bacterium with powerful antibiotics almost certainly does more harm than good. In the same manner, more people die *with* prostate cancer than *of* prostate cancer. Drastic treatments to eradicate every prostate cancer cell found may do more harm than good.

While the public perceives that cancer screening "saves lives," the truth is far more nuanced. Some cancer screening does save lives. Some cancer screening does not. Even then, looking only at the number of "cancers prevented" with screening is highly misleading because it considers only the positives. How many lives are harmed by screening? Suppose you find a tiny breast cancer that was not destined to progress any further. Finding it on screening tests leads to mastectomy, chemotherapy, and a lifetime of worry. You might have disfiguring surgery that leaves you with a swollen arm for the rest of your life. Chemotherapy increases the risk of heart failure and a future cancer. The risks involved with early screening and detection are very real, but the public rarely hears about them.

Without good evidence of the benefits of screening, and a better understanding of why it may fail, we must rely on the ancient medical guiding principle of *Primum non nocere*, which means, "First, do no harm." The perspective offered by today's cancer paradigm explains why many national agencies are beginning to scale back on the amount of screening they recommend.

22

DIETARY DETERMINANTS OF CANCER

AS IS NOW abundantly clear, cancer is not a rare disease but rather a common occurrence. Luckily, most cancers cause no problem and are discovered only incidentally, after death. Autopsy studies find unsuspected prostate cancer in 30 percent of men over the age of fifty;[1] 50 percent by age seventy;[2] and an astounding 80 percent by age ninety. If he lives long enough, every man can expect to develop prostate cancer. This holds true for other cancers, too. An estimated 11.2 percent of the adult population harbors thyroid cancer. Despite this high incidence, thyroid cancer only very rarely causes death. Colonoscopy screening studies find adenomas (a precancerous lesion) in almost half of the general population by age eighty.[3]

Because the seed of cancer is ever-present in all our cells, an important question is: why *don't* you get cancer? If it is not a seed problem, then it may be a soil problem. Diet is a hugely important determinant of progression because nutrient availability is inextricably linked to cell growth, particularly for cancer cells. Normal cells need both nutrients and growth factors to proliferate, but the growth signaling of cancer cells is always switched on, so the only limiting factor is nutrients.

An estimated 35 percent of cancers are attributed to diet/nutrition, making it the second most important determinant of cancer, trailing only tobacco smoking and dwarfing almost every other risk factor.[4] More specifically, excess body weight may be responsible for a large proportion of that risk.[5] The prevalence of most types of cancer has been slowly decreasing over time. Conspicuously, obesity-related cancers are increasing in incidence, making diet one of the most important preventative strategies at our disposal today.

While it's exciting to know that we have control at least one variable in our cancer risk, I'm afraid that many readers may find this chapter disappointing. I would love to reveal the "secret" to preventing or curing cancer. But cancer is not so simple—there is no miracle food or diet that can keep cancer at bay. Some preliminary studies suggest that certain foods may offer some protective benefit, but that's about it. For the most part, dietary prevention of cancer boils down to one key strategy: avoiding diseases of hyperinsulinemia, including obesity and type 2 diabetes.

WEIGHT LOSS

In Europe and North America, approximately 20 percent of incident cancer cases are attributable to obesity,[6] and for those who are overweight, intentional weight loss can reduce the risk of cancer death by 40 to 50 percent.[7] The clearest evidence is provided by studies on bariatric (weight loss) surgery. There are many ways to lose weight, but the data from these surgical studies is particularly instructive because of the clear date of intervention and the magnitude of the weight loss.

Several studies have documented a substantial benefit to intentional weight loss. A 2008 Canadian study found that deliberate weight loss from bariatric surgery reduces cancer risk by an estimated 78 percent.[8] In the Swedish Obese Subjects Study (SOS), a prospective, controlled intervention trial, bariatric surgery[9] was found to decrease cancer by an impressive 42 percent in women, but the rate was largely unchanged in men. The Utah Obesity Study of bariatric surgery also showed this unexpected difference between the sexes.[10] In women, total cancer incidence was reduced by 24 percent, but in men, once again, the rate was unchanged.

However, gastric bypass surgery carries its own risks, with some studies showing up to a doubling of the risk of colorectal cancer.[11] The damage and inflammation of surgery may stimulate hyperproliferation of the bowel mucosa and lead to cancer.[12] Given the cost, surgical risk, and potential increased risk of colorectal cancer, weight loss surgery cannot be widely recommended for cancer prevention. So, what other options are available?

Calorie restriction, defined as reduction of energy intake without malnutrition, was first shown to inhibit the growth of tumors in mice in 1909.[13] The cancer in mice given just enough food to survive barely grew at all. In mice that were allowed to eat as much as they wanted, the cancer grew the fastest. This protective effect is also found in monkeys, which showed a 50 percent reduction in cancer risk when fed a calorie-restricted diet.[14] Of course, translating results from animal studies into practical human solutions is problematic—maintaining drastic calorie restriction is extremely difficult for humans. I would guess that most people have tried a calorie-restricted diet at some point in their lives, and I would further guess that they are no longer on that diet.

Rather than restricting all calories, another strategy may be to reduce the most insulin-stimulating foods, such as sugar and refined carbohydrates. The nutrient sensor insulin/IGF-1 is a vital growth factor and plays a key role in causing obesity and type 2 diabetes. Research shows that reducing blood insulin may reduce the risk of cancer.[15] High insulin levels are also associated with a poor prognosis in cancer cases.[16] Two large prospective cohort studies, the Nurses' Health Study and the Health Professionals Follow-up Study, showed that a high dietary insulin load in cancer patients is associated with an increased risk of cancer recurrence and death.[17] For colon cancer patients, diets with the highest insulin effect more than doubled the risk of death compared to low-insulin diets (see Figure 22.1).[18]

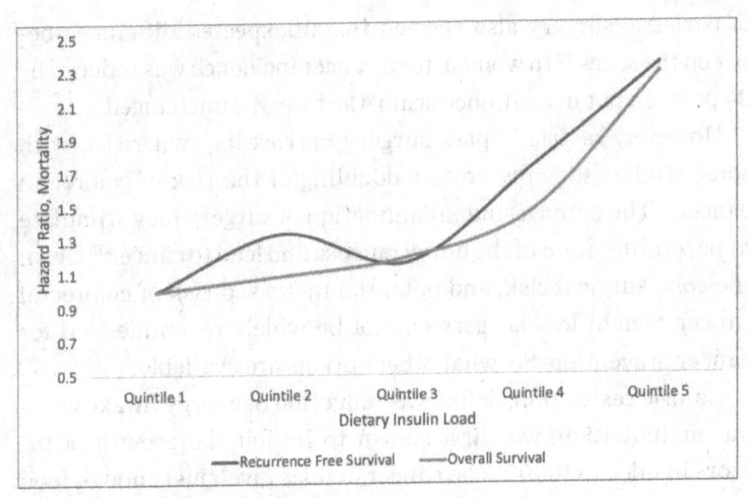

Figure 22.1

It is reasonable to hypothesize that diets targeted to reduce insulin effect may be beneficial for cancer, but definitive studies are lacking. Ketogenic diets, first described by Hans Krebs[19] in 1966, are high in fat, moderate in protein, and very low in carbohydrates. This forces the body to metabolize fat for fuel, instead of glucose. Keeping dietary carbohydrates low reduces both glucose and insulin. Ketogenic diets lower insulin, IGF-1, and mTOR without necessarily lowering calories—a health benefit for some, but the research is not yet there to support the benefit of a ketogenic diet in the treatment or prevention of cancer.

While maintaining a low insulin level may help to prevent cancer, the question of nutrition becomes much more complicated once cancer has progressed.

CANCER CACHEXIA

Cancer cachexia is the phenomenon of unintentional weight loss observed in patients with advanced disease and poor prognosis.

This syndrome can occur in people with other chronic diseases, too, such as chronic kidney disease, HIV, and tuberculosis. Weight loss is seen in 30 to 80 percent of cancer patients and tends to be progressive.[20] Generally, the more weight lost, the worse the cancer prognosis. Cachexia differs from normal weight loss because both body fat and muscle are lost.

Cachexia is not simply caused by loss of appetite due to the side effects of treatment. The specific mechanism for cancer cachexia is unknown, but it is likely due to the release of inflammatory cytokines—inflammatory signaling molecules—such as tumor necrosis factor (TNF) alpha. Weight loss due to lack of food (intentional or not) differs fundamentally from cancer cachexia. In anorexia nervosa, for example, after the first several days of fasting (not eating), more than 75 percent of the energy of metabolism is derived from body fat, sparing functional protein and muscle. By contrast, approximately equal amounts of muscle and fat are lost for energy production in cancer cachexia.[21] This results in the muscle wasting typical of cancer cachexia, which does not occur in fasting, unless it is extreme.

Intentional weight loss is usually accompanied by a slowing of the basal metabolic rate (BMR) to compensate for the reduced food availability. This does not happen in cancer cachexia, and BMR remains inappropriately high. Weight loss continues even as patients become more and more malnourished. Conceptually, cachexia is yet another mechanism that benefits the cancer at the organism's expense. When fat is metabolized for energy, molecules called ketone bodies are produced, which cancer cells have difficulty using. By stimulating the breakdown of muscle protein, amino acids are delivered to the liver and converted to glucose, which cancer cells love. Treatment of cancer cachexia is very difficult because simply eating more food does not reduce the inflammatory cytokines, so it does not prevent the muscle loss or the wasting syndrome. Even if weight is regained, it may be regained as fat, while muscle loss continues unabated.

So, while weight loss may be a useful strategy to prevent the progression of cancer, cancer cachexia, once advanced, limits the effect of diet on cancer treatment. Reducing glucose in an attempt to "starve" cancer is only modestly helpful because advanced cancer can break down other tissues to free the glucose it needs. Cancer may also metabolize amino acids such as glutamine, which are released during muscle breakdown. Dietary therapy must likely be combined with other treatments to be effective at this stage.

FASTING AND CANCER

Intermittent fasting is a promising nutritional approach for cancer prevention, as it protects against many of the risk factors such as obesity, type 2 diabetes,[22] and inflammation.[23] Low-carbohydrate diets reduce glucose and insulin, but not the other nutrient sensors, mTOR and AMPK. Fasting simultaneously reduces all the human nutrient sensors and most of the growth pathways, such as PI3K, mTOR, and IGF-1,[24] and also increases autophagy and mitophagy. One recent study found that women who fast for fewer than thirteen hours per night, despite having a lower BMI than other women in the study who fasted for that duration, had a 36 percent higher risk of recurrent breast cancer.[25]

Fasting during chemotherapy may also reduce the side effects of treatment while increasing efficacy. Chemotherapy targets cancer's rapidly proliferating cells; in the process, other normal but quick-growing cells, such as the hair follicles and the lining of the gastrointestinal system, also sustain damage. Fasting protects normal cells by putting them into a quiescent state, or maintenance mode, which may help mitigate chemotherapy's side effects of hair loss and nausea. Cancer cells do not enjoy this protective state because their genetic programming puts them into continuous growth mode.

In small clinical trials, patients had no difficulty fasting before and after chemotherapy,[26] and this seems to protect against side effects such as fatigue, weakness, and GI upset.[27] More important, fasting may also enhance the effectiveness of the chemotherapy.[28] In animal and cell line models, starvation conditions increased the effect of chemotherapy in fifteen of seventeen mammalian cancer cell lines. A reduction in side effects may allow a higher dosing of drug, resulting in increased killing of cancer cells.

CHEMOPREVENTION

The term *cancer chemoprevention* was introduced by the NIH in 1976 to denote foods, supplements, or drugs that may block progression of cancer. One of the most promising chemopreventive drugs is the old diabetes drug metformin. Studies have demonstrated that in type 2 diabetics, metformin may potentially reduce the risk of cancer by as much as 21 percent to 57 percent.[29] Specifically, long-term metformin use in women with type 2 diabetes was associated with a more than 50 percent reduction in breast cancer risk.[30] Metformin works by reducing glucose to the growing cancer without the pro-growth effects of the insulin/IGF-1/PI3K pathway. It also activates AMPK, an important nutrient sensor and growth pathway, which rapidly inhibits cellular protein synthesis and growth. Some research has indicated that this beneficial anticancer effect might extend to nondiabetics as well.[31]

The most widely studied natural food for chemoprevention is green tea, which contains high levels of chemical compounds called catechins.[32] Green tea contains much higher concentrations of catechins than black teas, with the catechins accounting for up to 30 percent of the green tea's dry weight and perhaps a higher percentage in cold-brewed green tea crystals. There are several potential health benefits to drinking green tea, although most of these clinical studies are quite small. Drinking green tea may help reduce some of the risk factors for cancer such as excess weight,[33] insulin resistance,[34] inflammation,[35] and type 2 diabetes.[36]

In 2000, researchers in Japan found that high consumption of green tea delayed the average age of cancer onset by 7.3 years[37] and reduced breast cancer recurrence.[38] Green tea extract supplements were shown to reduce the incidence of colorectal adenomas by over 50 percent in small pilot studies.[39] In prostate cancer, green tea extracts block the progression of high-grade precancerous lesions.[40] These studies are promising but highly preliminary, but green tea is one of the few chemopreventive tools that is a low-cost, natural food with no negative side effects.

At this point, the available scientific studies can make only the following firm recommendations:

- If you are overweight, lose weight; and
- Avoid or reverse type 2 diabetes.

Two other not-so-firm recommendations are:

- If you have type 2 diabetes, consider metformin; and
- Consider drinking more green tea.

The primary role of diet in cancer treatment is to reduce cancer progression rather than treat the disease. The mainstay of treatment is to reduce the growth factor availability, predominantly insulin, and avoid the hyperinsulinemic states of obesity and type 2 diabetes.

But just imagine how powerful changing the diet can be. Imagine if a Japanese woman in America could reduce her risk of breast cancer to that of a Japanese woman in Japan. Imagine if we could change our diets by eating natural foods, and reduce our risk of certain cancers so that, like the indigenous peoples of the Far North, we, too, were considered "immune" to cancer.

23

IMMUNOTHERAPY

I CAN BEAT Michael Jordan. I can beat Tiger Woods, too. *What's that?* you say. You may wonder if I've taken leave of my senses. Not at all. It's really very simple: I don't play them at basketball or golf, but instead challenge them to a contest on medical physiology. I'd be crazy to challenge Michael Jordan at basketball. I'd be crazy to challenge Tiger Woods at golf. As the ancient Chinese strategist Sun Tzu wrote in *The Art of War* in the fifth century BC, "In war, the way is to avoid what is strong and strike at what is weak."

How does this philosophy apply to cancer treatments? The previous paradigms of cancer all failed because they focused on attacking cancer's strengths, not its weaknesses.

Cancer paradigm 1.0 viewed cancer as a disease of excessive growth, and its core strength is that it grows and survives better than anything else in the world. We tried to kill it, but we were playing cancer at its own game. There were some notable successes, but the limits of this approach were quickly realized. We attacked the disease's main strength head-on. Sometimes we won, but too often, we lost. When cancer recurred, it became resistant to previous treatments and continued growing. Cancer evolved over time and space, but our treatments did not.

Cancer is not a mindless growing machine. It is a dynamic, evolving species bent on its own survival. Chemotherapy targets cell growth, arguably a living organism's most basic ability. But four billion years of evolution prepared cancer cells for this ultimate battle of survival. The growth pathways targeted by chemotherapy are likely the cancer cell's least vulnerable, most redundant capabilities.

Cancer paradigm 2.0 imagined that cancer was fundamentally a disease of randomly accumulated genetic mutations. Block this mutation (or at most two or three mutations), and cancer was cured. There were some great successes, but once again, the limits of this approach were quickly evident. What happened? Once again, we were attacking cancer at its strengths. Cancer is constantly mutating. So, we tried to devise ways to block those mutations. We were trying to beat Tiger Woods in a game of golf.

Block one pathway, and cancer will usually find an alternate route. The mutations that made cancer cells the ultimate survivors were not random, but were driven by the process of tumoral evolution. The evolutionary/ecological paradigm of cancer explains how some cancer treatments fail, but it can also point us in a new, more strategic direction. Cancer is an invasive species fighting for its very existence.

Luckily, we have evolved multipronged defenses against alien invaders. The logical strategy to win the war is to enhance our own innate defense, the immune system—which brings us to the most promising treatment of the last thirty years: immunotherapy.

COLEY'S TOXINS

In 1829, a woman suffering progressive breast cancer refused surgery. After battling the cancer for eighteen months, she was bedridden, cachectic, and close to death when she developed a high fever. The cancer was inflamed, so the physician made several incisions into the tumor to remove some fluid. Within eight days, the cancer shrank to one third its original size, and within four weeks, no trace of it remained. What happened? How did the infection *cure* her cancer?[1]

In 1867, German physician Wilhelm Busch cauterized an incurable tumor on a female patient's neck. During recovery, she lay next to a patient suffering a skin infection known as erysipelas, caused by streptococcus bacteria. The cancer patient soon contracted the same infection and developed a high fever, whereupon her tumor immediately began to shrink. Friedrich Fehleisen, another German physician, repeated this treatment in 1882, with some success. Cancer remission has also been documented in a patient infected with gas gangrene, an infection of the *Clostridium* species of bacteria.

The idea that our own immune system can fight cancer is not new. Spontaneous regression of malignant cancers is rare, but it occurs in approximately one in one hundred thousand cancer cases,[2] covering almost all different cancer types. Spontaneous regression is defined as the partial or complete disappearance of cancer in the absence of medical treatment. It is most commonly associated with an acute febrile illness, usually due to an infection or vaccination.

These fortuitous cures emboldened early physicians to exploit a primitive form of immunotherapy for otherwise incurable cancer. The ancient Egyptian physician Imhotep (ca. 2600 BC) suggested wrapping the cancerous area with a poultice, then incising it. Bacteria could enter the skin to cause an infection, which would occasionally cure the cancer. It was uncommon, but it was the patient's and physician's last, desperate hope.[3] Deliberate infections for the treatment of cancer continued to be prescribed well into the nineteenth century. Surgical wounds were knowingly left open to promote infection. Purulent sores were deliberately infected with septic dressings.[4]

In the 1880s, Fred Stein, a German immigrant to New York City, developed a fast-growing tumor in his neck. Despairing, doctors proclaimed this a hopeless case and advised him to put his affairs in order. He would soon succumb to the emperor of all maladies, they said. But fate would intervene. Stein developed facial erysipelas, and antibiotics had not yet been developed. Incredibly, his ramped-up immune system not only beat back the infection, but destroyed the cancer, too.

American surgeon Dr. William Coley tracked down Fred Stein in 1891. Intrigued by the body's intrinsic cancer-fighting ability, Coley spent the next decades trying to persuade the immune system to be a cancer killer.[5] He developed what could be considered the world's first attempt at a cancer vaccine. Coley inoculated patients with *Streptococcus pyogenes* bacteria to trigger erysipelas and provoke the immune system. Coley hoped this immune response would extend to the malignancy.[6] Results from such a primitive treatment were inconsistent. There were some great successes, tempered by some horrible failures. Intentionally infecting people in the era prior to antibiotics was not a particularly winning strategy.

But Coley was not deterred. The problem was not the efficacy, but the toxicity. He tinkered with his formula by adding other bacteria (*Serratia marcescens*) and inactivating them with heat before administration. This formula, now called Coley's toxins, was ultimately used to treat more than a thousand patients with inoperable cancer, including lymphomas, myelomas, carcinomas, and melanomas. Coley's toxin was injected directly into the tumor daily for one to two months before the dose was gradually tapered.

Some of the results were astounding. More than half of previously inoperable sarcomas showed complete remission, and the patient survived for more than five years. Even after twenty years, 21 percent of patients had no evidence of cancer. For its time, this was nothing short of miraculous. Interestingly, this treatment was used even for advanced disease, because the immune system could find and attack cancer anywhere in the body, even if it had spread.

Coley himself emphasized that causing a fever was paramount for inducing spontaneous regression. Coley's toxins were last used in China in 1980, where a patient with terminal liver cancer was treated over the course of thirty-four weeks. His symptoms disappeared completely. With the advent of chemotherapy and the genetic paradigm of cancer, Coley's toxins have been seen only in the history books, and for decades, the idea of cancer immunotherapy was pushed aside.

IMMUNE EDITING

The fact that some cancers could spontaneously regress implied that intrinsic forces within our own bodies could both prevent cancer and destroy it. In 1909, German scientist Paul Ehrlich proposed a radical new theory of cancer. Counter to the prevailing belief that it was relatively rare, Ehrlich conjectured that cancer cells are relatively common, but are prevented from doing more harm by an intrinsic host defense, now known to be the immune system.[7] Although he did not know it at the time, he was describing the concept of "immune surveillance," whereby the human immune system constantly identifies and eradicates evolving tumors.

This hypothesis was further refined in 1970 by Nobel Prize–winning immunologist Sir Frank Burnet,[8] who suggested that the genetic changes of malignancy were not rare. Burnet proposed that the immune system eliminates these dangerous cells as part of its routine surveillance to keep the body in good working order. Burnet wrote that a "small accumulation of tumor cells may develop and . . . provoke an effective immunological reaction with regression of the tumor and no clinical hint of its existence."[9] Cancer cells are developing constantly, but they are wiped out by our innate immune defenses. This original concept of immune surveillance has been extended further and now is called immunoediting, which consists of three stages: elimination (immune surveillance), equilibrium, and escape.

Elimination

Societies regularly employ specialized forces, like the police and drug enforcement agencies, to seek out and destroy disruptive elements in their midst. Similarly, the human immune system employs specialized cells to regularly patrol the entire body, on the lookout for disruptive elements like viruses, bacteria, and, yes, cancer cells. When they find damaged cells that are potentially cancerous, they kill them with great prejudice. This quickly eliminates the cancerous threat before it can spread.

Every cell in the body is estimated to experience more than twenty thousand DNA-damaging events every day.[10] Every day! Chronic sublethal cellular damage is not a rare event; it is an everyday occurrence. Common sources of damage include smoke, air pollution, viruses, bacteria, and radiation. Luckily, our cells have evolved potent DNA-repair pathways, but if the balance is tilted in favor of cellular damage, these repair mechanisms may be unable to cope. All damaged cells are potentially cancerous. At a certain level of damage, those cells are better removed via apoptosis or destroyed by the immune system.

Any weakening of immunity, such as with HIV infection or drugs, predisposes cells to succumbing to cancer. The immune system also degrades with age, which might help explain why cancer risk increases so significantly with age. If the immune system is not strong enough to completely eradicate the cancer cells, then we move from elimination of cancer to equilibrium.

Equilibrium

In 2004, a sixty-four-year-old man with pulmonary fibrosis underwent a single lung transplant and received the standard high-dose immune suppression to prevent organ rejection. A year later, he developed shortness of breath and cough. A right lung nodule was identified as a metastatic melanoma, as were several lymph nodes.

Cancer! This man had never in his life suffered from melanoma, so why was this melanoma so widespread in his lung now? A thorough examination of his skin revealed no evidence of melanoma then or ever. With increasing dread, the medical team asked for more information about the medical history of the lung donor.

The donor was a fifty-one-year-old woman who had died from trauma. Her medical history was unremarkable, but upon further investigation, a single problem was found. At the age of twenty-one, this woman had undergone surgery to remove a melanoma. She required no further treatment and had never suffered a relapse. The cancer had presumably been cured many, many years prior and did not preclude her from organ donation, even if the transplant team had known about it. At the time of organ harvest, her lungs looked completely normal and healthy.

Further DNA testing confirmed that the recipient's melanoma did indeed come from the lung donor. In this case, a microscopic sleeper cell of viable cancer lived on in the donor's lungs despite no clinical evidence of disease. The cancer had simply lain dormant, kept in check for thirty years by the woman's immune system. The cancer cells avoided eradication by her immune system, but were not strong enough to spread further. It was an uneasy stalemate. When her lung was transplanted into the immune-suppressed recipient, the delicate equilibrium between cancer and immune system swung decidedly in favor of cancer.[11] Within seven months of diagnosis, the man died of metastatic melanoma.

During the equilibrium stage of immune editing, pockets of cancer cells survive but cannot proliferate due to the suppression by the immune system. Latent cancers can lie dormant for years or decades. Cancer's efforts to grow are matched by the immune system's efforts to contain that growth. The pro- and anticancer forces are equally balanced.

Escape

With age, the immune system generally weakens and sometimes can no longer contain the cancer. In this struggle between the cancer, an invasive species, and the immune system, the advantage shifts to the cancer. It escapes repression by the immune system.

Untreated, the cancer will continue to grow and then metastasize. At this point, a boost to the immune system may well tip the balance in our favor. This is the promise of the next frontier in cancer medicine: immunotherapy.

IMMUNOTHERAPY

The Early Years

In 1929, physicians at John Hopkins noted that patients with tuberculosis (TB) seemed somehow protected against cancer. TB reduced the risk of cancer by almost 60 percent![12] Tuberculosis, endemic in many parts of the world, is caused by a slow-growing bacterium called *Mycobacterium tuberculosis*. It is resistant to most antibiotics, and even today, treatment revolves around isoniazid, a drug first discovered in 1912. The widespread and largely incurable nature of TB led to the establishment of many sanatoria in the nineteenth century where patients were quarantined. The lack of effective treatment spurred great interest in the development of the Bacillus Calmette–Guérin (BCG) vaccine in 1921, using the closely related *Mycobacterium bovis* bacterium.

Animal studies in the 1950s suggested that the BCG vaccine also protected against cancer. By 1976, studies proved that it effectively treated superficial bladder cancer in humans.[13] In 1990, BCG directly instilled into the bladder by cystoscope was approved by the FDA for the treatment of bladder cancer.

An astounding 71 percent of patients with early-stage bladder cancer respond to BCG treatment,[14] which represents today's first-line treatment of bladder cancer. How exactly this TB vaccine works for superficial bladder cancer was completely unknown. It worked, and that's about all we knew. BCG profoundly stimulates the immune system,[15] and this somehow improved recognition and subsequent destruction of the cancerous cells.

In 1992, there was another brief flurry of interest in cancer immunotherapy with the development of interleukin-2 (IL-2) treatment. IL-2 drove the T cells, an integral part of the immune system, into a frenzy, and cancer cells were killed in the crossfire. But because of the general activation of the immune system, IL-2 also caused a lot of side effects, such as fever, chills, nausea, and diarrhea. Ultimately, IL-2 treatment was deemed effective for only about 6 percent of melanoma patients, but about 2 percent of patients died from its side effects.[16] Fortunately, cancer immunotherapy has progressed far beyond these early, tentative steps.

Modern Immunotherapy

Nobel laureate Dr. James Allison harbored a deep personal vendetta against cancer, having lost his mother to lymphoma, his uncle to lung cancer, and his brother to prostate cancer. In 1978, he began researching how T cells attack tumors, becoming an early pioneer in cancer immunology long before it was considered a reputable field.

The human immune system contains many types of cells. The T cells are the professional killers, highly lethal cells designed to destroy pathogens. The body therefore maintains tight controls over these deadly weapons. T cells must kill sick or infected cells, but leave the normal ones alone. Unchecked, T cells could devastate the body. Autoimmune diseases, such as systemic lupus erythematosus and rheumatoid arthritis, are caused by an overresponsive immune system. The goal is to kill all the invaders but avoid friendly fire. To accomplish this, the immune system must accurately distinguish between "self" and "non-self" tissues. Like a nuclear missile, a healthy immune system must be highly lethal but tightly regulated. For this, we use both positive and negative controls.

To launch a nuclear missile, two keys must be simultaneously activated to reduce the chance of an errant launch. That's the positive control mechanism for activation, and for added protection, there's also a negative control, a "kill switch" to immediately abort the launch in case of an emergency. The kill switch makes many prominent appearances in Hollywood action movies, where the hero deactivates the missile about to incinerate a densely populated city with only a single second left. The human T cells work in the same manner. Two receptors must be simultaneously activated for a T cell to trigger. The T cell must detect both the tumor antigen and a second switch, known as the costimulatory receptor CD28.

At the time of Allison's research, nobody had yet suspected another layer of protection: the negative control, or kill switch. In the 1990s Allison was working with a newly described receptor known as cytotoxic T-lymphocyte-associated protein #4 (CTLA-4), which most researchers believed was a T cell activator. Allison's breakthrough was the recognition that CTLA-4 was not an activator switch but a kill switch. The possibility that T cells had a kill switch had never been previously considered.[17]

If both "go" signals were present, then the T cell would initiate Rambo mode and start destroying enemies, especially cancer cells. The kill switch, CTLA-4, acted as a checkpoint for the T cells. It was the final decision maker. If CTLA-4 was not engaged, the T cells would launch their nuclear assault. If the kill switch was hit, then the immune attack would shut down. Cancer cells avoid the highly lethal T cells by mimicking this kill switch.

So, if we could disable this kill switch, we could unleash the T cells' all-out assault on cancer cells. By 1996, Allison engineered a monoclonal antibody that could block CTLA-4 and become the world's first checkpoint inhibitor.[18] In one of his first animal experiments, he administered this new drug and watched in astonishment as the tumors completely melted away. The tumors in the mice that did not receive the antibody kept growing. Allison recalls, "It was the perfect experiment. 100 percent alive versus 100 percent dead."

This antibody became known as ipilimumab, and it was approved in 2011 by the FDA for the treatment of metastatic melanoma. It was the first drug of any type to improve survival in advanced melanoma and provided proof for the concept of cancer immunotherapy. More than 20 percent of patients with metastatic melanoma who received ipilimumab were still alive ten years later.[19] The results are even more amazing if you consider that ipilimumab is given for only three months. That sort of durable response is virtually unknown in oncology because of cancer's frustrating ability to evolve.

But CTLA-4 is not the only T cell kill switch in the human immune system. In 1992, working independently at Kyoto University, in Japan, Dr. Tasuku Honjo discovered another T cell kill switch called the programmed cell death protein 1 (PD-1). Normal healthy cells express PD-1 on their surface to protect them from immune attack. Fetal cells, for example, are covered in PD-1, which protects them from Mom's immune cells.

Cancer cells use the same trick, producing abundant PD-1 to disguise themselves as normal cells and protect against attack from the immune system—a wolf in sheep's clothing, a classic survival ploy. An antibody that blocked PD-1 would release the kill switch, allowing T cells to target the uncloaked cancer cells. By 2012, a second class of checkpoint inhibitors against PD-1 proved their efficacy in human cancer and were approved by the FDA in 2014. These drugs were effective against a wide variety of tumors, including melanoma, lung cancer, and kidney cancer. Honjo and Allison would share the 2018 Nobel Prize in Physiology or Medicine for having "established an entirely new principle for cancer therapy."[20] Combining the PD-1- and CTLA-4-blocking antibodies may offer an even more effective treatment.[21]

Another promising immunotherapy is a technology called adoptive T cell transfer. In this treatment, a patient's own T cells are extracted and then grown in a lab. A cancer-targeting system called a chimeric antigen receptor (CAR-T) is attached to the T cells, which are transferred back into the patient. These activated, lethal T cells home in on the patient's specific cancer like a precision-guided missile. The first two CAR-T treatments received FDA approval in 2017: tisagenlecleucel, for the treatment of leukemia, and axicabtagene ciloleucel, for the treatment of lymphoma.[22] CAR-T is more a delivery platform than a drug, because new chimeric antigens can be attached to a patient's T cells. Theoretically, CAR-T could provide the opportunity to target any cancer.

Immunotherapy has several inherent advantages over conventional treatments. First, cancer is always in a state of dynamic evolution with its environment. A drug attacks a static target and does not evolve. Thus, cancer easily stickhandles around the drug to develop resistance and render these treatments ineffective over time. The boosted immune system is a dynamic system that can better keep pace with cancer's moves. The immune system can adjust and evolve alongside the cancer.

Second, the immune system has a memory, so it may prevent recurrence. When we get vaccinated for measles as a child, our immune system remembers this virus and provides lifetime protection. In the same manner, the boosted immune system allowed some melanoma patients extended survival possibly due to this memory effect.

Third, immunotherapy has fewer side effects than standard chemotherapies because the immune system is a targeted treatment. Conventional chemotherapy is a toxic treatment, designed to kill cancer cells slightly faster than regular cells. Immunotherapy is not intrinsically toxic, except to cells that are identified by the body as alien.

Fourth, immunotherapy is a systemic treatment, which is crucial because cancer is a systemic disease. Metastasis occurs early in the disease process, so a systemic therapy can treat *potential* micrometastases throughout the body. The immune system can lock on and destroy the cancer cells and does not need to be manually targeted in the same way as local treatments like surgery and radiation. The systemic nature of treatment also means that immunotherapy may be effective even very late in the disease process, after the cancer has metastasized. Even from the inception of immunotherapy, Coley observed that this systemic effect can potentially benefit late-stage cancer patients.

Affordability, though, is not one of the advantages of immunotherapy. Given the exorbitant price tag of these treatments, many providers question the feasibility of using these advanced drugs. In single-payer systems, the possible cost of saving a few lives must be balanced with other things—like more hospital beds, more nursing care, or more home care. This is not an easy question, and is one that lies beyond the scope of this book, but it is an issue that will soon be at the forefront of health care discussions.

THE ABSCOPAL EFFECT

In 2008, a thirty-three-year-old woman with surgically treated melanoma underwent a PET scan that lit up a new two-centimeter lung nodule. The cancer had returned. Part of her lung was removed, and she started chemotherapy and maintenance immunotherapy with ipilimumab, which put the melanoma in remission, but only temporarily. By 2010, new metastatic lesions were noted in her spleen, chest, and the lining of the lung (pleura). A painful lesion near her spine was treated with fractionated radiation. As expected, the spinal metastasis shrank, but remarkably, so did the metastases in her spleen and chest, even though they were outside the radiation field.[23] How could a local therapy (radiation) result in a systemic response to the cancer that had invaded so many areas of her body?

This 2012 case report in the *New England Journal of Medicine* rediscovered a phenomenon known as the "abscopal effect," first observed in 1973. The term *abscopal*[24] is derived from the Latin prefix *ab-*, meaning "far," and *-scopus*,[25] meaning "target." An abscopal effect has a consequence far away from its intended target. Radiotherapy burns away cancerous tissue or any other tissue that gets in its way. Sometimes, unexpectedly, nonirradiated metastatic lesions regress far from the site of treatment.

Radiation usually affects only the cells within the irradiated area. But in a very small percentage of cases, cancer cells outside and even far away from that area have also responded to the treatment. This kind of outcome has historically been rare—at least, until the age of immunotherapy. The medical literature from 1969 to 2018 reported ninety-four cases of the abscopal effect, but strikingly, half of those reports came from the last six years, the era of modern immunotherapy.[26] When combined with immunotherapy, radiation may induce an effective systemic anti-tumor response that far exceeds the expected benefit of the two treatments administered alone. A recent study found this abscopal effect in an astounding 27 percent of patients with metastatic solid tumors treated with immunotherapy and radiation.[27] The widespread use of immunotherapy has changed the abscopal effect from an atypical phenomenon to one that could potentially benefit more than a quarter of all cancer patients.

As strange as this phenomenon may seem at first glance, the evolutionary paradigm of cancer can help us understand why the abscopal effect takes place. Radiation damages cellular DNA and causes necrosis. This uncontrolled cell death splatters cellular parts around the tissues like raw eggs dropped onto a sidewalk. The DNA, normally tightly contained within the nucleus, suddenly becomes exposed, and this highly inflammatory state attracts immune cells to clean up the mess. In addition, the immune system is primed to seek and destroy similar-looking cells.

But cancer cells are protected because they cloak themselves using PD-1 and CTLA-4 and do not elicit a sufficient immune reaction. When patients receive *both* radiation and immunotherapy, the immune system is not only activated but also primed to kill the uncloaked cancer cells specifically. This synergy is responsible for the abscopal effect. The local radiation-induced cell damage acts much like a vaccine, aiming the activated immune system at the DNA target like a guided missile. But using the right dosing regimen is crucial.

Normally, exposed DNA is cleaned up by a cellular enzyme called TREX1. This enzyme, imaginatively named for the dinosaur species *Tyrannosaurus rex*, voraciously gobbles up any stray DNA lying around to prevent further problems. Higher radiation doses activate TREX1, which destroys stray DNA, prevents immune system activation, and reduces the abscopal effect.[28] A smaller radiation dose spread out over time (fractionated), staying low enough to avoid TREX1 activation, may be more effective at producing the abscopal effect.[29]

In 2019, the first small controlled human trial targeting the abscopal effect was published.[30] All patients were treated with immunotherapy and randomized to additional radiation or not. Those patients receiving the radiation doubled their objective response rate, and median overall survival more than doubled from 7.6 months to 15.9 months. Because of the small numbers of patients enrolled, these results were not statistically significant, but they are encouraging nonetheless.

The difference between an average athlete and a Hall of Famer is the latter's ability to make those around him better. Immunotherapy represents the future of cancer medicine, not only because it is effective by itself, but also because it makes other, older treatments better.

ADAPTIVE THERAPY

The biggest problem with standard cancer treatments is not that they don't kill cancer cells; they do that just fine. The problem is that cancer develops resistance. Chemotherapy, radiation, and hormonal treatments all kill cancer cells, but they also exert a natural selective pressure favoring resistance. These are all inherently double-edged weapons, with the potential both to cure and to kill. The evolutionary paradigm of cancer brings up an important question: is it even necessary to eradicate cancer, or is it enough simply to control its population?

In 1989, cancer researcher Robert Gatenby became fascinated by the idea of tumoral evolution. Cancer cells, he reasoned, must compete for resources. Mathematical models used since the 1920s have described how populations grew under harsh conditions. For example, Lotka-Volterra equations model the growth of a population of snowshoe hares and the lynx that feed upon them. Gatenby applied such equations to populations not of hares but of cancer cells,[31] pioneering the field of mathematical oncology.

The population of an invasive species involves dispersal, proliferation, migration, and evolution, exactly as cancer does. For example, a crop-eating pest grows quickly when food is easily available. Pesticides destroy the pests, but inevitably, resistance develops even against the most potent pesticides (e.g., the infamous DDT). Cancer cells, too, develop resistance to the most potent chemotherapies. Successful eradication of widespread pests is rare because the pesticides act as a natural selective pressure favoring drug resistance. These resistant pests face reduced competition and therefore thrive.

Suppose you spray one billion locusts with pesticide to reduce that population by 99.9 percent to one million. Those one million remaining locusts now have no competition for food and begin to expand their population exponentially. Eventually, you end up with a billion pesticide-resistant locusts. Cancer cells are no different. You can kill 99.9 percent of cancer cells with chemotherapy, but those surviving cells face reduced competition and therefore plenty of resources to thrive. Also, the new population of cancer cells will be treatment-resistant.

The standard treatment mantra of chemotherapy is to administer the maximally tolerated dosage (MTD)—that is, give as much chemotherapy as humanly possible without killing the patient. When Gatenby mathematically modeled that strategy, resistance developed virtually every time, resulting in eventual treatment failure.[32]

In 2014, Gatenby tested a promising new strategy, called adaptive therapy, based on his mathematical modeling. If the "treat to kill" strategy doesn't work for metastatic cancer, he reasoned, then perhaps a "treat to contain" strategy could work. Rather than carpet bombing the cancer with MTD, he selectively gave chemotherapy only above a certain level of cancer activity, trying to manage, rather than eradicate, the cancer. As they trickled in, the results of the pilot study were astounding. Adaptive therapy—that is, the use of less than half the dose of the expensive chemotherapy drug—improved survival by 64 percent.[33]

Resistant strains of cancer must devote more resources toward maintaining their resistance. Without the drug to act as natural selection pressure, the drug-resistant strains are disadvantaged, using precious resources to maintain a relatively useless drug-resistance trait. While these results are preliminary, they do highlight the types of innovative research that can be done using new cancer paradigms. Perhaps in some cases, we can do better by controlling rather than eradicating cancer. Sometimes you score more points in the game by making easy layups instead of always going in for the slam dunk.

CONCLUSION

Immunotherapy, the abscopal effect, and adaptive therapy are examples of new cancer strategies revealed by the evolutionary paradigm of cancer. The technology behind immunotherapy is revolutionary, and the future is bright, but it has not yet arrived. Despite the number of drugs approved by the FDA between 2002 and 2014, the improvement in overall survival for solid tumors is just 2.1 months.[34] Still, for the first time in decades, there is reason to be optimistic.

Applying the lessons of evolutionary biology to our understanding of cancer brings new hope for the future of treatment. Will we turn the corner on cancer? Only time will tell, but our new understanding of our ancient foe promises a light at the end of this dark, dark tunnel.

EPILOGUE

CANCER IS FAR and away medicine's deepest mystery. Medical science has long unraveled the causes of many of the other diseases that afflict us. Infections are caused by bacteria, viruses, and fungi. Blockages in arteries cause heart disease, strokes, and peripheral vascular disease.

Cystic fibrosis is a genetic disease. Gout is caused by excess uric acid. Among the common diseases, cancer stands alone. What causes it? Why does it exist? What the heck *is* it?

We've moved through three great paradigms in our understanding of cancer. Cancer paradigm 1.0 considered cancer as a disease of excessive growth. Cancer paradigm 2.0 considered cancer as a disease of random genetic mutations that caused excessive growth. Both paradigms advanced our understanding of cancer's story, but they fell short. By pursuing the mystery of cancer's genesis past the beginnings of humanity and right to the edge of multicellular life, cancer paradigm 3.0 has revealed fresh insight into this fascinating foe.

The seed of cancer lies in all cells of all multicellular life. Cancer is an atavism, a reversion to an earlier genetic playbook brought on by the struggle for a cell to survive (transformation). Whether that seed flourishes depends upon the environment (the soil). The most important aspect of progression are the growth pathways of the body, which are also the nutrient-sensing pathways.

Diseases of growth are diseases of metabolism. Diseases of metabolism are diseases of growth. Cancer is a disease of evolution and ecology. While there undoubtedly is yet more to discover, this new paradigm represents a huge leap forward.

These fresh new insights into cancer bring fresh new treatments. We are finally seeing decreasing rates of cancer and cancer deaths, as we understand both the benefits and limitations of screening programs. We are honing our weapons of mass cellular destruction with greater precision as we recognize their double-edged nature. Maybe we don't need to always eradicate cancer cells. We are developing new systemic, immune-mediated weapons that will hunt down cancer cells and kill them wherever they may hide.

But one massive new obstacle has arisen. The burgeoning obesity crisis has increased the rates of the obesity-related cancers, including breast cancer and colorectal cancer. Where most cancers are gradually decreasing in prevalence, these cancers are increasing. But there is reason for optimism yet. Nutrition is our primary weapon against obesity-related cancers. By making changes to our diet, we can help to reduce the threat.

A NEW HOPE

Cancer medicine has been stuck in neutral for decades as the scientific and medical communities have slowly worked their way through each paradigm. But I'm hopeful for the future because our new understanding can drive progress in ways not previously imagined. Cancer is a disease unlike anything else we face in medicine. The story of cancer is stranger than science fiction, and it took some forceful insights from an astrobiologist to guide us toward the correct path.

The recognition of a new paradigm of cancer medicine means that for the first time in decades, we stand a chance at making real progress in our decades-long war on cancer. A new hope arises. A new dawn breaks.

To all those whose lives have been touched by cancer, from researchers to doctors to patients to families, I hope this book has helped to shed some light on this deepest of medical mysteries.

NOTES

Chapter 1: Trench Warfare

1. "Adult Obesity Prevalence Maps," Centers for Disease Control and Prevention, updated October 29, 2019, https://www.cdc.gov/obesity/data/prevalence-maps.html.
2. Max Frankel, "Protracted War on Cancer," *New York Times*, June 12, 1981, https://www.nytimes.com/1981/06/12/opinion/protracted-war-on-cancer.html.
3. J. C. Bailar III and E. M. Smith, "Progress Against Cancer?," *New England Journal of Medicine* 314, no. 19 (May 8, 1986): 1226–32.
4. Barron H. Lerner, "John Bailar's Righteous Attack on the 'War on Cancer,'" Slate, January 12, 2017, https://slate.com/technology/2017/01/john-bailar-reminded-us-of-the-value-of-evidence.html.
5. Clifton Leaf, *The Truth in Small Doses* (New York: Simon & Schuster, 2013), 25.
6. J. C. Bailar III and H. L. Gornik, "Cancer Undefeated," *New England Journal of Medicine* 336, no. 22 (May 29, 1997): 1569–74.
7. Gina Kolata, "Advances Elusive in the Drive to Cure Cancer," *New York Times*, April 23, 2009, https://www.nytimes.com/2009/04/24/health/policy/24cancer.html.
8. Alexander Nazaryan, "World War Cancer," *New Yorker*, June 30, 2013, https://www.newyorker.com/tech/annals-of-technology/world-war-cancer.
9. James D. Watson, "To Fight Cancer, Know the Enemy," *New York Times*, August 5, 2009, https://www.nytimes.com/2009/08/06/opinion/06watson.html.
10. David Chan, "Where Do the Millions of Cancer Research Dollars Go Every Year?," Slate, February 7, 2013, https://slate.com/human-interest/2013/02/where-do-the-millions-of-cancer-research-dollars-go-every-year.html.
11. J. R. Johnson et al., "End Points and United States Food and Drug Administration Approval of Oncology Drugs," *Journal of Clinical Oncology* 21, no. 7 (April 1, 2003): 1404–11.

Chapter 2: The History of Cancer

1. Siddhartha Mukherjee, *The Emperor of All Maladies* (New York: Simon & Schuster, 2010), 6.
2. "Anesthesia Death Rates Improve over 50 Years," CBC, September 21, 2012, https://www.cbc.ca/news/health/anesthesia-death-rates-improve-over-50-years-1.1200837.

3. E. H. Grubbe, "Priority in the Therapeutic Use of X-rays," *Radiology* 21 (1933): 156–62.
4. M. A. Cleaves, "Radium: With a Preliminary Note on Radium Rays in the Treatment of Cancer," *Medical Record* 64 (1903): 601–6.
5. E. B. Krumbhaar, "Role of the Blood and the Bone Marrow in Certain Forms of Gas Poisoning: I. Peripheral Blood Changes and Their Significance," *JAMA*, 72 (1919): 39–41.
6. I. Berenblum et al., "The Modifying Influence of Dichloroethyl Sulphide on the Induction of Tumours in Mice by Tar," *Journal of Pathology and Bacteriology* 32 (1929): 424–34.
7. Sarah Hazell, "Mustard Gas—from the Great War to Frontline Chemotherapy," Cancer Research UK, August 27, 2014, https://scienceblog.cancerresearchuk.org/2014/08/27/mustard-gas-from-the-great-war-to-frontline-chemotherapy/.
8. R. J. Papac, "Origins of Cancer Therapy," *Yale Journal of Biology and Medicine* 74 (2001): 391–98.
9. S. Farber et al., "Temporary Remissions in Acute Leukemia in Children Produced by Folic Acid Antagonist, 4-aminopteroyl-glutamic Acid (Aminopterin)," *New England Journal of Medicine* 238 (1948): 787–93.
10. M. C. Li, R. Hertz, and D. M. Bergenstal, "Therapy of Choriocarcinoma and Related Trophoblastic Tumors with Folic Acid and Purine Antagonists," *New England Journal of Medicine* 259 (1958): 66–74.
11. E. J. Freireich, M. Karon, and E. Frei III, "Quadruple Combination Therapy (VAMP) for Acute Lymphocytic Leukemia of Childhood," *Proceedings of the American Association for Cancer Research* 5 (1964): 20.
12. V. T. DeVita, A. A. Serpick, and P. P. Carbone, "Combination Chemotherapy in the Treatment of Advanced Hodgkin's Disease," *Annals of Internal Medicine* 73 (1970): 881–95.

Chapter 3: What Is Cancer?

1. Letter from Charles Darwin to J. D. Hooker, August 1, 1857, DCP-LETT-2130, Darwin Correspondence Project, https://www.darwinproject.ac.uk/letter/?docId=letters/DCP-LETT-2130.xml.
2. D. Hanahan and R. A. Weinberg, "The Hallmarks of Cancer," *Cell* 100, no. 1 (January 2000): 57–70.
3. D. Hanahan and R. A. Weinberg, "Hallmarks of Cancer: The Next Generation," *Cell* 144, no. 5 (March 4, 2011): 646–74, doi: 10.1016/j.cell.2011.02.013.
4. A. G. Renehan et al., "What Is Apoptosis, and Why Is It important?," *BMJ* 322 (2001): 1536–38.
5. J. F. Kerr, A. H. Wyllie, and A. R. Currie, "Apoptosis: A Basic Biological Phenomenon with Wide-Ranging Implications in Tissue Kinetics," *British Journal of Cancer* 26, no. 4 (August 1972): 239–57.
6. J. W. Shay et al., "Hayflick, His Limit, and Cellular Ageing," *Nature Reviews Molecular Cell Biology* 1, no. 1 (October 2000): 72–76, doi: 10.1038/35036093.

7. G. Watts, "Leonard Hayflick and the Limits of Ageing," *Lancet* 377, no. 9783 (June 18, 2011): 2075, doi: 10.1016/S0140-6736(11)60908-2.
8. Robin McKie, "Henrietta Lacks's Cells Were Priceless, but Her Family Can't Afford a Hospital," *Guardian*, April 3, 2010, https://www.theguardian.com/world/2010/apr/04/henrietta-lacks-cancer-cells.
9. O. Warburg, F. Wind, and E. J. Negelein, "The Metabolism of Tumors in the Body," *General Physiology* 8, no. 6 (March 7, 1927): 519–30.

Chapter 4: Carcinogens

1. "Cancer," Mayo Clinic, December 12, 2018, https://www.mayoclinic.org/diseases-conditions/cancer/symptoms-causes/syc-20370588.
2. D. E. Redmond, "Tobacco and Cancer: The First Clinical Report, 1761," *New England Journal of Medicine* 282, no. 1 (January 1, 1970): 18–23.
3. J. R. Brown and J. L. Thornton, "Percivall Pott (1714–1788) and Chimney Sweepers' Cancer of the Scrotum," *British Journal of Industrial Medicine* 14, no. 1 (January 1957): 68–70.
4. Daniel King, "History of Asbestos," Asbestos.com and the Mesothelioma Center, https://www.asbestos.com/asbestos/history/.
5. K. M. Lynch and W. A. Smith, "Pulmonary Asbestosis III: Carcinoma of the Lung in Asbesto-silicosis," *American Journal of Cancer* 24 (1935): 56–64.
6. B. I. Casteman, "Asbestos and Cancer: History and Public Policy," *British Journal of Industrial Medicine* 48 (1991): 427–32.
7. J. C. McDonald and A. D. McDonald, "Epidemiology of Mesothelioma, " in D. Liddell and K. Miller, eds., *Mineral Fibers and Health* (Boca Raton, FL: CRC Press, 1991).
8. M. Albin et al., "Asbestos and Cancer: An Overview of Current Trends in Europe," *Environmental Health Perspectives* 107, Suppl. 2 (1999): 289–98, https://ehp.niehs.nih.gov/doi/10.1289/ehp.99107s2289.
9. J. LaDou, "The Asbestos Cancer Epidemic," *Journal of Environmental Health Perspectives* 112, no. 3 (2004): 285–90.
10. "Agents Classified by the IARC Monographs, Volumes 1–127," International Agency for Research on Cancer, June 26, 2020, https://monographs.iarc.fr/agents-classified-by-the-iarc/.
11. D. J. Shah et al., "Radiation-induced Cancer: A Modern View," *British Journal of Radiology* 85, no. 1020 (2012): e1166–73.
12. K. Ozasa et al., "Studies of the Mortality of Atomic Bomb Survivors: Report 14, 1950–2003: An Overview of Cancer and Noncancer Diseases," *Radiation Research* 177 (2012): 229–43.
13. B. R. Jordan, "The Hiroshima/Nagasaki Survivor Studies: Discrepancies between Results and General Perception," *Genetics* 203, no. 4 (2016): 1505–12.
14. J. F. Kerr et al., "Apoptosis: A Basic Biological Phenomenon with Wide-Ranging Implications in Tissue Kinetics," *British Journal of Cancer* 26 (1972): 239–57.

Chapter 5: Cancer Goes Viral

1. D. A. Burkitt, "Sarcoma Involving the Jaws in African Children," *British Journal of Surgery* 46 (1958): 218–23, doi: 10.1002/bjs.18004619704.
2. I. Magrath, "Denis Burkitt and the African Lymphoma," *Ecancermedicalscience* 3, no. 159 (2009): 159, doi: 10.3332ecancer.2009.159.
3. M. A. Epstein, B. G. Achong, and Y. M. Barr, "Virus Particles in Cultured Lymphoblasts from Burkitt's Lymphoma," *Lancet* 1 (1964): 702–3, doi: 10.1016/S0140-6736(64)91524-7.
4. J. S. Pagano et al., "Infectious Agents and Cancer: Criteria for a Causal Relation," *Seminars in Cancer Biology* 14 (2004): 453–71.
5. D. P. Burkitt, "Etiology of Burkitt's Lymphoma: An Alternative Hypothesis to a Vectored Virus," *Journal of the National Cancer Institute* 42 (1969): 19–28.
6. Magrath, "Dennis Burkitt and the African Lymphoma."
7. M. L. K. Chua et al., "Nasopharyngeal Carcinoma," *Lancet* 387, no. 10022 (2016): 1012–24.
8. F. Petersson, "Nasopharyngeal Carcinoma: A Review," *Seminars in Diagnostic Pathology* 32, no. 1 (2015): 54–73.
9. E. Chang and H.-O. Adami, "The Enigmatic Epidemiology of Nasopharyngeal Carcinoma," *Cancer Epidemiol, Biomarkers and Prevention* 15 (2006): 1765–77.
10. Nicholas Wade, "Special Virus Cancer Program: Travails of a Biological Moon Shot," *Science* 174 (December 24, 1971): 1306–11.
11. Harold M. Schmeck Jr., "National Cancer Institute Reorganizing 10-Year-Old Viral Research Program," *New York Times*, June 19, 1974, https://www.nytimes.com/1974/06/19/archives/national-cancer-institute-reorganizing-10yearold-viral-research.html.
12. T. M. Block et al., "A Historical Perspective on the Discovery and Elucidation of the Hepatitis B Virus," *Antiviral Research* 131 (July 2016): 109–23, doi: 10.1016/j.antiviral.2016.04.012.
13. R. P. Beasley et al., "Hepatocellular Carcinoma and Hepatitis B Virus: A Prospective Study of 22 707 Men in Taiwan," *Lancet* 8256 (1981): 1129–33.
14. V. Vedham et al., "Early-Life Exposures to Infectious Agents and Later Cancer Development," *Cancer Medicine* 4, no. 12 (2015): 1908–22.
15. H. J. Alter et al., "Posttransfusion Hepatitis After Exclusion of Commercial and Hepatitis-B Antigen-Positive Donors," *Annals of Internal Medicine* 77, no. 5 (1972): 691–99.
16. Zosia Chustecka, "Nobel-Winning Discovery of HPV–Cervical Cancer Link Already Having an Impact on Medicine," Medscape Medical News, October 16, 2008, http://www.vch.ca/Documents/public-health-nobel-winning-hpv.pdf.
17. J. M. Walboomers et al., "Human Papillomavirus Is a Necessary Cause of Invasive Cervical Cancer Worldwide," *Journal of Pathology* 189, no. 1 (September 1999): 12–19.
18. L. Torre et al., "Global Cancer Statistics," *CA: A Cancer Journal for Clinicians* 65 (2012): 87–108.

19. M. Arbyn et al., "Prophylactic Vaccination Against Human Papillomaviruses to Prevent Cervical Cancer and Its Precursors," Cochrane Database of Systematic Reviews 5, Article No. CD009069, 2018, doi: 10.1002/14651858.CD009069.pub3.
20. K. D. Crew and A. I. Neugut, "Epidemiology of Gastric Cancer," *World Journal of Gastroenterology* 12, no. 3 (January 21, 2006): 354–62.
21. B. Linz et al., "An African Origin for the Intimate Association between Humans and *Helicobacter pylori*," *Nature* 445 (2007): 915–18.
22. Pamela Weintraub, "The Doctor Who Drank Infectious Broth, Gave Himself an Ulcer, and Solved a Medical Mystery," *Discover*, April 8, 2010, http://discovermagazine.com/2010/mar/07-dr-drank-broth-gave-ulcer-solved-medical-mystery.
23. S. Suerbaum and P. Michetti P., "*Helicobacter pylori* Infection," *New England Journal of Medicine* 347 (2002): 1175–86.
24. H. S. Youn et al., "Pathogenesis and Prevention of Stomach Cancer," *Journal of Korean Medical Science* 11 (1996): 373–85.
25. A. M. Nomura, G. N. Stemmermann, and P. H. Chyou, "Gastric Cancer among the Japanese in Hawaii," *Japanese Journal of Cancer Research* 86 (1995): 916–23.
26. D. M. Parkin et al., "Global Cancer Statistics, 2002," *CA: A Cancer Journal for Clinicians* 55 (2005): 74–108.
27. R. Mera et al., "Long-term Follow-up of Patients Treated for *Helicobacter pylori* Infection," *Gut* 54 (2005): 1536–40.
28. M. E. Stolte et al., 2002. "*Helicobacter* and Gastric MALT Lymphoma," *Gut* 50, Suppl. 3 (2002): III19–III24.
29. R. M. Peek Jr. and J. E. Crabtree, "*Helicobacter* Infection and Gastric Neoplasia," *Journal of Pathology* 208 (2006): 233–48.
30. D. Parkin, "The Global Health Burden of Infection-associated Cancers in the Year 2002," *International Journal of Cancer* 118 (2006): 3030–44.

Chapter 6: The Somatic Mutation Theory

1. J. Gayon, "From Mendel to Epigenetics: History of Genetics," *Comptes Rendus Biologies* 339 (2016): 225–30.
2. T. Boveri, "Über mehrpolige Mitosen als Mittel zur Analyse des Zellkerns," *Verh. D. Phys. Med. Ges. Würzberg N. F.* 35 (1902): 67–90.
3. A. Balmain, "Cancer Genetics: From Boveri and Mendel to Microarrays," *Nature Reviews* 1 (2001): 77–82.
4. K. Bister, "Discovery of Oncogenes: The Advent of Molecular Cancer Research," *PNAS* 112, no. 50 (2015): 15259–60.
5. L. Chin et al., "*P53* Deficiency Rescues the Adverse Effects of Telomere Loss and Cooperates with Telomere Dysfunction to Accelerate Carcinogenesis," *Cell* 97 (1999): 527–38.
6. "Known and Probable Human Carcinogens," American Cancer Society, last updated August 14, 2019, https://www.cancer.org/cancer/cancer-causes/general-info/known-and-probable-human-carcinogens.html.

7. A. Balmain and I. B. Pragnell, "Mouse Skin Carcinomas Induced *in vivo* by Chemical Carcinogens Have a Transforming Harvey-ras Oncogene," *Nature* 303 (1983): 72–74.
8. "Age and Cancer Risk," NIH National Cancer Institute, April 29, 2015, https://www.cancer.gov/about-cancer/causes-prevention/risk/age.
9. P. Nowell and D. Hungerford, "A Minute Chromosome in Human Chronic Granulocytic Leukemia," abstract, *Science* 132 (1960): 1497.
10. E. H. Romond et al., "Trastuzumab Plus Adjuvant Chemotherapy for Operable HER2 Positive Breast Cancer," *New England Journal of Medicine* 353 (2005): 1673–84.

Chapter 7: Cancer's Procrustean Bed

1. P. Lichtenstein, "Environmental and Heritable Factors in the Causation of Cancer," *New England Journal of Medicine* 343 (2000): 78–85.
2. M. C. King, J. H. Marks, and J. B. Mandell, "Breast and Ovarian Cancer Risks Due to Inherited Mutations in *BRCA1* and *BRCA2*," *Science* 302, no. 5645 (2003): 643–46.
3. L. A. Mucci et al., "Familial Risk and Heritability of Cancer among Twins in Nordic Countries," *JAMA* 315, no. 1 (January 5, 2006): 68–76.
4. A. R. David and M. R. Zimmerman, "Cancer: An Old Disease, a New Disease or Something in Between?," *Nature Reviews Cancer* 10 (2010): 728–33.
5. James W. Hampton, "Cancer Prevention and Control in American Indians/ Alaska Natives," *American Indian Culture and Research Journal* 16, no. 3 (1992): 41–49.
6. M. L. Sievers and J. R. Fisher, "Cancer in North American Indians: Environment versus Heredity," *American Journal of Public Health* 73 (1983): 485–87; T. K. Young and J. W. Frank, "Cancer Surveillance in a Remote Indian Population in Northwestern Ontario," *American Journal of Public Health* 73 (1983): 515–20.
7. I. M. Rabinowitch, "Clinical and Other Observations on Canadian Eskimos in the Western Arctic," *Canadian Medical Association Journal* 34 (1936): 487.
8. O. Schafer et al., "The Changing Pattern of Neoplastic Disease in Canadian Eskimos," *Canadian Medical Association Journal* 112 (1975): 1399–1404.
9. F. S. Fellows, "Mortality in the Native Races of the Territory of Alaska, with Special Reference to Tuberculosis," *Public Health Reports (1896–1970)* 49, no. 9 (March 2, 1934): 289–98.
10. J. T. Friborg and M. Melbye, "Cancer Patterns in Inuit Populations," *Lancet Oncology* 9, no. 9 (2008): 892–900.
11. R. G. Ziegler et al., "Migration Patterns and Breast Cancer Risk in Asian-American Women," *Journal of the National Cancer Institute* 85, no. 22 (November 17, 1993): 1819–27.
12. J. Peto, "Cancer Epidemiology in the Last Century and the Next Decade," *Nature* 411, no. 6835 (May 17, 2001): 390–95.
13. Andrew Pollack, "Huge Genome Project Is Proposed to Fight Cancer," *New York Times*, March 28, 2005, https://www.nytimes.com/2005/03/28/health/huge-genome-project-is-proposed-to-fight-cancer.html.

14. Pollack, "Huge Genome Project Is Proposed to Fight Cancer."
15. G. L. G. Miklos, "The Human Cancer Genome Project—One More Misstep in the War on Cancer," *Nature Biotechnology* 23 (2005): 535–37.
16. "NIH Completes In-depth Genomic Analysis of 33 Cancer Types," NIH National Cancer Institute, April 5, 2018, https://www.cancer.gov/news-events/press-releases/2018/tcga-pancancer-atlas.
17. T. Sjöblom et al., "The Consensus Coding Sequences of Human Breast and Colorectal Cancers," *Science* 314, no. 5797 (2006): 268–74.
18. Heidi Ledford, "End of Cancer-Genome Project Prompts Rethink," *Nature Magazine*, January 5, 2015, https://www.scientificamerican.com/article/end-of-cancer-genome-project-prompts-rethink/.
19. L. D. Wood et al., "The Genomic Landscapes of Human Breast and Colorectal Cancers," *Science* 318, no. 5853 (November 16, 2007): 1108–13.
20. S. Yachida et al., "Distant Metastasis Occurs Late During the Genetic Evolution of Pancreatic Cancer," *Nature* 467, no. 7319 (October 28, 2010): 1114–17.
21. Bert Vogelstein et al., "Cancer Genome Landscapes," *Science* 339, no. 6127 (March 29, 2013): 1546–58, doi: 10.1126/science.1235122.
22. B. Pereira et al., "The Somatic Mutation Profiles of 2,433 Breast Cancers Refines Their Genomic and Transcriptomic Landscapes," *Nature Communications* 10, no. 7 (2016): 11479, doi: 10.1038/ncomms11479.
23. C. Greenman et al., "Patterns of Somatic Mutation in Human Cancer Genomes," *Nature* 446, no. 7132 (March 8, 2007): 153–58.
24. Vogelstein et al., "Cancer Genome Landscapes," 1546–58.
25. Yachida et al., "Distant Metastasis Occurs Late During the Genetic Evolution of Pancreatic Cancer," 1114–17.
26. L. A. Loeb et al., "A Mutator Phenotype in Cancer," *Cancer Research* 61, no. 8 (April 15, 2001): 3230–39.

Chapter 8: The Denominator Problem

1. D. Humpherys et al., "Abnormal Gene Expression in Cloned Mice Derived from Embryonic Stem Cell and Cumulus Cell Nuclei," *Proceedings of the National Academy of Sciences* 99, no. 20 (October 1, 2002): 12889–94.
2. K. B. Jacobs et al., "Detectable Clonal Mosaicism and Its Relationship to Aging and Cancer," *Nature Genetics* 44, no. 6 (May 6, 2012): 651–58.
3. Carl Zimmer, "Researchers Explore a Cancer Paradox," *New York Times*, October 18, 2018, https://www.nytimes.com/2018/10/18/science/cancer-genetic-mutations.html.
4. I. Martincorena et al., "Somatic Mutant Clones Colonize the Human Esophagus with Age," *Science* 362, no. 6417 (October 18, 2018): 911–17, doi: 10.1126/science.aau3879.
5. A. G. Renehan et al., "The Prevalence and Characteristics of Colorectal Neoplasia in Acromegaly," *Journal of Clinical Endocrinology and Metabolism* 85, no. 9 (September 2000): 3417–24.
6. C. A. Sheldon et al., "Incidental Carcinoma of the Prostate: A Review of the Literature and Critical Reappraisal of Classification," *Journal of Urology* 124, no. 5 (November 1980): 626–31.

7. W. C. Hahn and R. A. Weinberg, "Mechanisms of Disease: Rules for Making Human Tumor Cells," *New England Journal of Medicine* 347 (2002): 1593–1603.
8. T. Sjoblom et al., "The Consensus Coding Sequences of Human Breast and Colorectal Cancers," *Science* 314 (2006): 268–74.
9. D. L. Stoler et al., "The Onset and Extent of Genomic Instability in Sporadic Colorectal Tumor Progression," *Proceedings of the National Academy of Sciences* 96, no. 26 (1999): 15121–26.

Chapter 9: A False Dawn

1. H. Bower et al., "Life Expectancy of Patients with Chronic Myeloid Leukemia Approaches the Life Expectancy of the General Population," *Journal of Clinical Oncology* 34, no. 24 (August 20, 2016): 2851–57, doi: 10.1200/JCO.2015.66.2866.
2. J. Elliott et al., "ALK Inhibitors for Non-Small Cell Lung Cancer: A Systematic Review and Network Meta-analysis," *PLoS One* 19, no. 15 (February 19, 2020): e0229179, doi: 10.1371/journal.pone.0229179.
3. "Crizotinib," GoodRx.com, https://www.goodrx.com/crizotinib.
4. D. M. Hyman et al., "Implementing Genome-driven Oncology," *Cell* 168, no. 4 (2017): 584–99.
5. Charles Ornstein and Katie Thomas, "Top Cancer Researcher Fails to Disclose Corporate Financial Ties in Major Research Journals," *New York Times*, September 8, 2018, https://www.nytimes.com/2018/09/08/health/jose-baselga-cancer-memorial-sloan-kettering.html.
6. I. F. Tannock and J. A. Hickman, "Limits to Personalized Cancer Medicine," *New England Journal of Medicine* 375 (2016): 1289–94.
7. V. Prasad, "Perspective: The Precision Oncology Illusion," *Nature* 537 (2016): S63.
8. F. Meric-Bernstam et al., "Feasibility of Large-Scale Genomic Testing to Facilitate Enrollment onto Genomically Matched Clinical Trials," *Journal of Clinical Oncology* 33, no. 25 (September 1, 2015): 2753–65.
9. K. T. Flaherty, et al. "NCI-Molecular Analysis for Therapy Choice. Interim Analysis Results." https://www.allianceforclinicaltrialsinoncology.org/main/cmsfile?cmsPath=/Public/Annual%20Meeting/files/CommunityOncology-NCI-Molecular%20Analysis.pdf.
10. J. Marquart, E. Y. Chen, and V. Prasad, "Estimation of the Percentage of US Patients with Cancer Who Benefit from Genome-driven Oncology," *JAMA Oncology* 4, no. 8 (2018): 1093–98.
11. D. S. Echt et al., "Mortality and Morbidity in Patients Receiving Encainide, Flecanide, or Placebo: The Cardiac Arrhythmia Suppression Trial," *New England Journal of Medicine* 324 (1991): 781–88.
12. C. M. Booth and E. A. Eisenhauer, "Progression-free Survival: Meaningful or Simply Measurable?," *Journal of Clinical Oncology* 30 (2012): 1030–33.
13. V. Prasad et al., "A Systematic Review of Trial-Level Meta-Analyses Measuring the Strength of Association between Surrogate End-Points and Overall Survival in Oncology," *European Journal of Cancer* 106

(2019): 196–211, doi: 10.1016/j.ejca.2018.11.012; Prasad et al., "The Strength of Association between Surrogate End Points and Survival in Oncology: A Systematic Review of Trial-Level Meta-analyses," *JAMA Internal Medicine* 175, no. 8 (2015): 1389–98, doi: 10.1001/jamainternmed. 2015.2829.

14. R. Kemp and V. Prasad, "Surrogate Endpoints in Oncology: When Are They Acceptable for Regulatory and Clinical Decisions, and Are They Currently Overused?," *BMC Medicine* 15, no. 1 (2017): 134.
15. J. Puthumana et al., "Clinical Trial Evidence Supporting FDA Approval of Drugs Granted Breakthrough Therapy Designation," *JAMA* 320, no. 3 (2018): 301–3.
16. E. Y. Chen et al., "An Overview of Cancer Drugs Approved by the US Food and Drug Administration Based on the Surrogate End Point of Response Rate," *JAMA Internal Medicine*, doi: 10.1001/jamainternmed.2019.0583.
17. B. Gyawali et al., "Assessment of the Clinical Benefit of Cancer Drugs Receiving Accelerated Approval," *JAMA Internal Medicine*, doi: 10.1001/jamainternmed.2019.0462.
18. K. Miller et al., "Paclitaxel plus Bevacizumab versus Paclitaxel Alone for Metastatic Breast Cancer," *New England Journal of Medicine* 357 (2007): 2666–76.
19. R. B. D'Agostino Sr., "Changing End Points in Breast-Cancer Drug Approval: The Avastin Story," *New England Journal of Medicine* 365, no. 2 (2011): e2.
20. Roxanne Nelson, "FDA Approves Everolimus for Advanced Breast Cancer," Medscape, July 20, 2012, https://www.medscape.com/viewarticle/767862.
21. M. Piccart et al., "Everolimus plus Exemestane for Hormone-Receptor-Positive, Human Epidermal Growth Factor Receptor-2-Negative Advanced Breast Cancer: Overall Survival Results from BOLERO-2," *Annals of Oncology* 25, no. 12 (2014): 2357–62.
22. V. Prasad et al., "The Strength of Association between Surrogate End Points and Survival in Oncology."
23. V. Prasad and S. Mailankody, "Research and Development Spending to Bring a Single Cancer Drug to Market and Revenues after Approval," *JAMA Internal Medicine* 177, no. 11 (2017): 1569–75, doi: 10.1001/jamainternmed.2017.3601.
24. E. Y. Chen et al., "Estimation of Study Time Reduction Using Surrogate End Points Rather than Overall Survival in Oncology Clinical Trials," *JAMA Internal Medicine* 179, no. 5 (2019): doi: 10.1001/jamainternmed.2018.8351.
25. T. Fojo et al., "Unintended Consequences of Expensive Cancer Therapeutics—The Pursuit of Marginal Indications and a Me-Too Mentality that Stifles Innovation and Creativity: The John Conley Lecture," *JAMA Otolaryngology Head and Neck Surgery* 140, no. 12 (2014): 1225–36, doi: 10.1001/jamaoto.2014.1570.
26. D. K. Tayapongsak et al., "Use of Word 'Unprecedented' in the Media Coverage of Cancer Drugs: Do 'Unprecedented' Drugs Live Up to the Hype?," *Journal of Cancer Policy* 14 (2017): 16–20.

27. M. V. Abola and V. Prasad, "The Use of Superlatives in Cancer Research," *JAMA Oncology* 2, no. 1 (2016): 139–41.
28. T. Rupp and D. Zuckerman, "Quality of Life, Overall Survival, and Costs of Cancer Drugs Approved Based on Surrogate Endpoints," *JAMA Internal Medicine* 177, no. 2 (2017): 276–77, doi: 10.1001/jamainternmed.2016.7761.
29. Carolyn Y. Johnson, "This Drug Is Defying a Rare Form of Leukemia—and It Keeps Getting Pricier," *Washington Post*, March 9, 2016, https://www.washingtonpost.com/business/this-drug-is-defying-a-rare-form-of-leukemia--and-it-keeps-getting-pricier/2016/03/09/4fff8102-c571-11e5-a4aa-f25866ba0dc6_story.html.
30. Prasad and Mailankody, "Research and Development Spending to Bring a Single Cancer Drug to Market and Revenues After Approval," 1569–75.
31. N. Gordon et al., "Trajectories of Injectable Cancer Drug Costs After Launch in the United States," *Journal of Clinical Oncology* 36, no. 4 (February 1, 2018): 319–25, doi: 10.1200/JCO.2016.72.2124.
32. V. Prasad, K. De Jesus, and S. Mailankody, "The High Price of Anticancer Drugs: Origins, Implications and Barriers, Solutions," *Nature Reviews Clinical Oncology* 14, no. 6 (2017): 381–90.
33. https://www.igeahub.com/2018/05/28/10-best-selling-drugs-2018-oncology/.
34. Alex Philippidis, "The Top 15 Best-Selling Drugs of 2017," GEN, March 12, 2018, https://www.genengnews.com/the-lists/the-top-15-best-selling-drugs-of-2017/77901068.
35. S. Singhal et al., "Antitumor Activity of Thalidomide in Refractory Multiple Myeloma," *New England Journal of Medicine* 341, no. 21 (1999): 1565–71.
36. Geeta Anand, "How Drug's Rebirth as Treatment for Cancer Fueled Price Rises," *Wall Street Journal*, November 15, 2004, https://www.wsj.com/articles/SB110047032850873523.
37. Hagop Kantarjian et al., "High Cancer Drug Prices in the United States: Reasons and Proposed Solutions," *Journal of Oncology Practice* 10, no. 4 (2014): e208–e211.
38. P. J. Neumann et al., "Updating Cost-effectiveness: The Curious Resilience of the $50,000-per-QALY Threshold," *New England Journal of Medicine* 371, no. 9 (August 2014): 28796–97, doi: 10.1056/NEJMp1405158.
39. Centers for Disease Control and Prevention, "Part V: Cost-Effectiveness Analysis," *Five-Part Webcast on Economic Evaluation*, April 26, 2017, https://www.cdc.gov/dhdsp/programs/spha/economic_evaluation/docs/podcast_v.pdf.
40. D. A. Goldstein, "Cost-effectiveness Analysis of Regorafenib for Metastatic Colorectal Cancer," *Journal of Clinical Oncology* 33, no. 32 (November 10, 2015): 3727–32, doi: 10.1200/JCO.2015.61.9569.
41. S. Mailankody and V. Prasad, "Five Years of Cancer Drug Approvals: Innovation, Efficacy, and Costs," *JAMA Oncology* 1, no. 4 (2015): 539–40.
42. Lorie Konish, "This Is the Real Reason Most Americans File for Bankruptcy," CNBC, February 11, 2019, https://www.cnbc.com/2019/02/11/this-is-the-real-reason-most-americans-file-for-bankruptcy.html.

Chapter 10: The Seed and the Soil

1. S. Paget, "The Distribution of Secondary Growths in Cancer of the Breast," *Lancet* 1 (1889): 99–101.
2. M. Esteller "Cancer Epigenomics: DNA Methylomes and Histone-Modification Maps," *Nature Reviews Genetics* 8, no. 4 (April 2007): 286–98.
3. L. J. C. Rush et al., "Novel Methylation Targets in De Novo Acute Myeloid Leukemia with Prevalence of Chromosome 11 Loci," *Blood* 97, no. 10 (2001): 3226–33, doi: 10.1182/blood.V97.10.3226.
4. A. D. Beggs et al., "Whole-genome Methylation Analysis of Benign and Malignant Colorectal Tumours," *Journal of Pathology* 229, no. 5 (April 2013): 697–704, doi: 10.1002/path.4132.
5. I. Martincorena et al., "Somatic Mutant Clones Colonize the Human Esophagus with Age," *Science* (October 18 2018): eaau3879, doi: 10.1126/science.aau3879.
6. Alaina G. Levine, "NIH Recruits Physicists to Battle Cancer," APS Physics, March 2010, https://www.aps.org/publications/apsnews/201003/nih.cfm.
7. R. A. Weinberg, "Coming Full Circle: From Endless Complexity to Simplicity and Back Again," *Cell* 157, no. 1 (March 27, 2014): 267–71, doi: 10.1016/j.cell.2014.03.004.
8. Paula Davies, "Cancer: The Beat of an Ancient Drum?," *Guardian*, April 25, 2011, https://www.theguardian.com/commentisfree/2011/apr/25/cancer-evolution-ancient-toolkit-genes.
9. Jessica Wapner, "A New Theory on Cancer: What We Know about How It Starts Could All Be Wrong," *Newsweek*, July 17, 2017, https://www.newsweek.com/2017/07/28/cancer-evolution-cells-637632.html.
10. Wapner, "A New Theory on Cancer."

Chapter 11: The Origins of Life and the Origins of Cancer

1. Michael Marshall, "Timeline: The Evolution of Life," *New Scientist*, July 14, 2009, https://www.newscientist.com/article/dn17453-timeline-the-evolution-of-life.
2. Leyland Cecco, "Rising Fame: Experts Herald Canadian Woman's 120-year-old Sourdough Starter," *Guardian*, May 14, 2018, https://www.theguardian.com/world/2018/may/14/ione-christensen-canada-yukon-sourdough-starter-yeast.
3. A. H. Yona et al., "Chromosomal Duplication Is a Transient Evolutionary Solution to Stress," *Proceedings of the National Academy of Sciences USA* 109, no. 51 (2012): 21010–15.
4. L. Cisneros et al., "Ancient Genes Establish Stress-induced Mutation as a Hallmark of Cancer," *PLoS One* 12, no. 4 (2017): e0176258.

Chapter 12: Tumoral Evolution

1. G. H. Heppner, "Tumor Heterogeneity," *Cancer Research* 44 (1984): 2259–65.

2. Cancer Genome Atlas Research Network, "Integrated Genomic Analyses of Ovarian Carcinoma," *Nature* 474 (2011): 609–15.
3. N. Navin et al., "Tumour Evolution Inferred by Single-cell Sequencing," *Nature* 472 (2011): 90–94.
4. S. Nik-Zainal et al., "Mutational Processes Molding the Genomes of 21 Breast Cancers," *Cell* 149 (2012): 979–93.
5. Charles Swanton, "Intratumor Heterogeneity: Evolution through Space and Time," *Cancer Research* 72, no. 19 (October 2012): 4875–82.
6. S. P. Shah et al., "Mutational Evolution in a Lobular Breast Tumour Profiled at Single Nucleotide Resolution," *Nature* 461 (2009): 809–13.
7. M. Gerlinger et al., "Intratumor Heterogeneity and Branched Evolution Revealed by Multiregion Sequencing," *New England Journal of Medicine* 366 (2012): 883–92.
8. L. Bai and W. G. Zhu, "*p53*: Structure, Function and Therapeutic Applications," *Journal of Molecular Cancer* 2, no. 4 (2006): 141–53.
9. A. Kamb, S. Wee, and C. Lengauer, "Why Is Cancer Drug Discovery So Difficult?," *Nature Reviews Drug Discovery* 6 (2007): 115–20.
10. L. M. Byrd et al., "Better Life Expectancy in Women with BRCA2 Compared with BRCA1 Mutations Is Attributable to Lower Frequency and Later Onset of Ovarian Cancer," *Cancer Epidemiology, Biomarkers and Prevention* 17, no. 6 (June 2008): 1535–42, doi: 10.1158/1055-9965.EPI-07-2792.
11. Wapner, "A New Theory on Cancer."
12. Helen Roberts, "Boy Born with 'Tail' Loses 'Monkey God' Status after It's Removed so He Can Walk," *Mirror*, July 3, 2015, https://www.mirror.co.uk/news/world-news/boy-born-tail-loses-monkey-5993397.

Chapter 13: Cancerous Transformation

1. T. Domazet-Lošo and D. Tautz, "Phylostratigraphic Tracking of Cancer Genes Suggests a Link to the Emergence of Multicellularity in Metazoa," *BMC Biology* 8, no. 66 (2010), https://doi.org/10.1186/1741-7007-8-66 PMID: 20492640.
2. A. S. Trigos et al., "Altered Interactions between Unicellular and Multicellular Genes Drive Hallmarks of Transformation in a Diverse Range of Solid Tumors," *Proceedings of the National Academy of Sciences USA* 114 (2017): 6406–11.
3. "COSMIC Release v90," Sanger Institute, September 5, 2019, https://cosmic-blog.sanger.ac.uk/cosmic-release-v90/.
4. L. Cisneros et al., "Ancient Genes Establish Stress-induced Mutation as a Hallmark of Cancer," *PLoS One* 12, no. 4 (2017): e0176258, https://doi.org/10.1371/journal.pone.0176258.
5. H. Chen et al., "The Reverse Evolution from Multi-cellularity to Unicellularity During Carcinogenesis," *Nature Communications* 6 (2015): 6367.
6. Vogelstein et al., "Cancer Genome Landscapes," 1546–58.
7. Chen et al., "The Reverse Evolution from Multi-cellularity to Unicellularity During Carcinogenesis," 6367.

8. H. Chen and X. He, "The Convergent Cancer Evolution Toward a Single Cellular Destination," *Molecular Biology and Evolution* 33, no. 1 (2016): 4–12, doi: 10.1093/molbev/msv212.
9. M. Vincent, "Cancer: A De-repression of a Default Survival Program Common to All Cells?," *Bioessays* 34 (2011): 72–82.
10. H. F. Dvorak, "Tumors: Wounds that Do Not Heal—Similarities between Tumor Stroma Generation and Wound Healing," *New England Journal of Medicine* 315, no. 26 (December 25, 1986): 1650–59.
11. L. Simonato et al., "Lung Cancer and Cigarette Smoking in Europe: An Update of Risk Estimates and an Assessment of Inter-country Heterogeneity," *International Journal of Cancer* 91, no. 6 (March 15, 2001): 876–87.
12. S. Bhat et al., "Risk of Malignant Progression in Barrett's Esophagus Patients: Results from a Large Population-Based Study," *Journal of the National Cancer Institute* 103, no. 13 (July 6, 2001): 1049–57.
13. L. A. Anderson et al., "Risk Factors for Barrett's Oesophagus and Oesophageal Adenocarcinoma: Results from the FINBAR Study," *World Journal of Gastroenterology* 13, no. 10 (March 14, 2007): 1585–94.
14. N. D. Walter et al., "Wound Healing after Trauma May Predispose to Lung Cancer Metastasis," *American Journal of Respiratory Cell and Molecular Biology* 44 (2011): 591–96.
15. R. P. DerHagopian et al., "Inflammatory Oncotaxis," *JAMA* 240, no. 4 (1978): 374–75.
16. L. M. Burt et al., "Risk of Secondary Malignancies after Radiation Therapy for Breast Cancer: Comprehensive Results," *Breast* 35 (October 2017): 122–29, doi: 10.1016/j.breast.2017.07.004.
17. M. Faurschou et al., "Malignancies in Wegener's Granulomatosis: Incidence and Relation to Cyclophosphamide Therapy in a Cohort of 293 Patients," *Journal of Rheumatology* 35, no. 1 (January 2008): 100–105.
18. J. A. Baltus et al., "The Occurrence of Malignancies in Patients with Rheumatoid Arthritis Treated with Cyclophosphamide: A Controlled Retrospective Follow-up," *Annals of the Rheumatic Diseases* 42, no. 4 (August 1983): 368–73.
19. H. Welch and W. C. Black, "Using Autopsy Series to Estimate the Disease 'Reservoir' for Ductal Carcinoma In Situ of the Breast: How Much More Breast Cancer Can We Find?," *Annals of Internal Medicine* 127 (1997): 1023–28.
20. "It Is Not the Strongest of the Species that Survives But the Most Adaptable," Quote Investigator, May 4, 2014, https://quoteinvestigator.com/2014/05/04/adapt/.

Chapter 14: Nutrition and Cancer

1. R. Doll and R. J. Peto, "The Causes of Cancer: Quantitative Estimates of Avoidable Risks of Cancer in the United States Today," *National Cancer Institute* 66, no. 6 (June 1981): 1191–308.

2. W. J. Blot and R. E. Tarone, "Doll and Peto's Quantitative Estimates of Cancer Risks: Holding Generally True for 35 Years," *Journal of the National Cancer Institute* 107, no. 4 (2015): djv044.
3. D. P. Burkitt, "Some Diseases Characteristic of Modern Western Civilization," *British Medical Journal* 1, no. 274 (1973): 274–78.
4. G. E. McKeown-Eyssen et al., "A Randomized Trial of a Low-Fat, High-Fiber Diet in the Recurrence of Colorectal Polyps," *Journal of Clinical Epidemiology* 47 (1994): 525–36.
5. R. MacLennan et al., "Randomized Trial of Intake of Fat, Fiber, and Beta Carotene to Prevent Colorectal Adenomas: The Australian Polyp Prevention Project," *Journal of the National Cancer Institute* 87 (1995): 1760–66.
6. C. S. Fuchs et al., "Dietary Fiber and the Risk of Colorectal Cancer and Adenoma in Women," *New England Journal of Medicine* 340, no. 3 (January 21, 1999): 169–76.
7. A. Schatzkin et al., "Lack of Effect of a Low-Fat, High-Fiber Diet on the Recurrence of Colorectal Adenomas," *New England Journal of Medicine* 342 (2000): 1149–55.
8. B. V. Howard et al., "Low-fat Dietary Pattern and Risk of Cardiovascular Disease: The Women's Health Initiative Randomized Controlled Dietary Modification Trial," *JAMA* 295, no. 6 (February 8, 2006): 655–66.
9. R. L. Prentice et al., "Low-Fat Dietary Pattern and Risk of Invasive Breast Cancer: The Women's Health Initiative Randomized Controlled Dietary Modification Trial," *JAMA* 295, no. 6 (February 8, 2006): 629–42.
10. S. A. A. Beresford et al., "Low-Fat Dietary Pattern and Risk of Colorectal Cancer: The Women's Health Initiative Randomized Controlled Dietary Modification Trial," *JAMA* 295, no. 6 (2006):643–54, doi: 10.1001/jama.295.6.643.
11. Alpha-Tocopherol, Beta Carotene Cancer Prevention Study Group, "The Effect of Vitamin E and Beta Carotene on the Incidence of Lung Cancer and Other Cancers in Male Smokers," *New England Journal of Medicine* 330, no. 13 (1994): 1029–35.
12. G. S. Omenn et al., "Effects of a Combination of Beta Carotene and Vitamin A on Lung Cancer and Cardiovascular Disease," *New England Journal of Medicine* 334, no. 18 (1996): 1150–55.
13. E. Lonn et al., "Heart Outcomes Prevention Evaluation (HOPE) 2 Investigators: Homocysteine Lowering with Folic Acid and B Vitamins in Vascular Disease," *New England Journal of Medicine* 354 (2006): 1567–77.
14. B. F. Cole et al., "Folic Acid for the Prevention of Colorectal Adenomas: A Randomized Clinical Trial," *JAMA* 297, no. 21 (June 6, 2007): 2351–59.
15. C. B. Ambrosone et al., "Dietary Supplement Use During Chemotherapy and Survival Outcomes of Patients with Breast Cancer Enrolled in a Cooperative Group Clinical Trial (SWOG S0221)," *Journal of Clinical Oncology* (December 19, 2019): JCO1901203, doi: 10.1200/JCO.19.01203.

16. K. H. Bønaa et al., "Homocysteine Lowering and Cardiovascular Events after Acute Myocardial Infarction," *New England Journal of Medicine* 354, no. 15 (2006): 1578–88.
17. M. Ebbing et al., "Mortality and Cardiovascular Events in Patients Treated with Homocysteine-lowering B Vitamins after Coronary Angiography: A Randomized Controlled Trial," *JAMA* 300, no. 7 (2008): 795–804.
18. M. Ebbing et al., "Cancer Incidence and Mortality after Treatment with Folic Acid and Vitamin B12," *JAMA* 302, no. 19 (November 18, 2009): 2119–26, doi: 10.1001/jama.2009.1622.
19. S. Faber et al., "The Action of Pteroylglutamic Conjugates on Man," *Science* 106 (1947): 619–21.
20. E. Cameron and L. Pauling, "Ascorbic Acid and the Glycosaminoglycans: An Orthomolecular Approach to Cancer and Other Diseases," *Oncology* 27, no. 2 (1973): 181–92
21. B. Lee et al., "Efficacy of Vitamin C Supplements in Prevention of Cancer: A Meta-Analysis of Randomized Controlled Trials," *Korean Journal of Family Medicine* 36, no. 6 (November 2015): 278–85.
22. S. Peller and C. S. Stephenson, "Skin Irritation and Cancer in the United States Navy," *American Journal of Medical Sciences* 194 (1937): 326–33.
23. F. L. Apperly, "The Relation of Solar Radiation to Cancer Mortality in North America," *Cancer Research* 1 (1941): 191–95.
24. C. F. Garland and F. C. Garland, "Do Sunlight and Vitamin D Reduce the Likelihood of Colon Cancer?," *International Journal of Epidemiology* 9 (1980): 227–31; W. B. Grant, "An Estimate of Premature Cancer Mortality in the US Due to Inadequate Doses of Solar Ultraviolet-B Radiation," *Cancer* 94 (2002): 1867–75.
25. N. Keum and E. Giovannucci, "Vitamin D Supplements and Cancer Incidence and Mortality: A Meta-analysis," *British Journal of Cancer* 111 (2014): 976–80.
26. K. K. Deeb, D. L. Trump, and C. S. Johnson, "Vitamin D Signalling Pathways in Cancer: Potential for Anticancer Therapeutics," *Nature Reviews Cancer* 7 (2007): 684–700, doi: 10.1038/nrc2196.
27. D. Feldman et al., "The Role of Vitamin D in Reducing Cancer Risk and Progression," *Nature Reviews Cancer* 14 (2014): 342–57.
28. M. L. Melamed et al., "25-hydroxyvitamin D Levels and the Risk of Mortality in the General Population," *Archives of Internal Medicine* 168 (2008): 1629–37, doi: 10.1001/archinte.168.15.1629.
29. J. E. Manson et al., "Vitamin D Supplements and Prevention of Cancer and Cardiovascular Disease," *New England Journal of Medicine* 380, no. 1 (January 3, 2019): 33–44, doi: 10.1056/NEJMoa1809944; J. E. Manson et al., "Marine n-3 Fatty Acids and Prevention of Cardiovascular Disease and Cancer," *New England Journal of Medicine* 380, no. 1 (January 3, 2019): 23–32, doi: 10.1056/NEJMoa1811403.
30. R. Scragg et al., "Monthly High-Dose Vitamin D Supplementation and Cancer Risk: A Post Hoc Analysis of the Vitamin D Assessment

Randomized Clinical Trial," *JAMA Oncology* 4, 11 (November 2018): e182178, doi: 10.1001/jamaoncol.2018.2178.
31. J. Ju et al., "Cancer Preventive Activities of Tocopherols and Tocotrienols. *Carcinogenesis* 31, no. 4 (April 2010): 533–42; S. Mahabir et al., "Dietary Alpha-, Beta-, Gamma- and Delta-tocopherols in Lung Cancer Risk," *International Journal of Cancer* 123 (2008): 1173–80.
32. I. M. Lee et al., "Vitamin E in the Primary Prevention of Cardiovascular Disease and Cancer: The Women's Health Study: A Randomized Controlled Trial," *JAMA* 294 (2005): 56–65.
33. D. Albanes et al., "Alpha-Tocopherol and Beta-carotene Supplements and Lung Cancer Incidence in the Alpha-tocopherol, Beta-carotene Cancer Prevention Study: Effects of Base-line Characteristics and Study Compliance," *Journal of the National Cancer Institute* 88 (1996): 1560–70.
34. J. M. Gaziano et al., "Vitamins E and C in the Prevention of Prostate and Total Cancer in Men: The Physicians' Health Study II Randomized Controlled Trial," *JAMA* 301 (2009): 52–62.
35. S. M. Lippman et al., "Effect of Selenium and Vitamin E on Risk of Prostate Cancer and Other Cancers: The Selenium and Vitamin E Cancer Prevention Trial (SELECT)," *JAMA* 301 (2009): 39–51.
36. E. A. Klein et al., "Vitamin E and the Risk of Prostate Cancer: The Selenium and Vitamin E Cancer Prevention Trial (SELECT)," *JAMA* 306 (2011): 1549–56.
37. B. Lauby-Secretan et al., "Body Fatness and Cancer: Viewpoint of the IARC Working Group," *New England Journal of Medicine* 375 (2016): 794–98.
38. E. E. Calle et al., "Overweight, Obesity, and Mortality from Cancer in a Prospectively Studied Cohort of U.S. Adults," *New England Journal of Medicine* 348, 17 (April 24, 2003): 1625–38.
39. C. Brooke Steele et al., "Vital Signs: Trends in Incidence of Cancers Associated with Overweight and Obesity—United States, 2005–2014," *Morbidity and Mortality Weekly Report* 66 (2017): 1052–58. https://www.cdc.gov/mmwr/volumes/66/wr/mm6639e1.htm.
40. Lauby-Secretan et al., "Body Fatness and Cancer," 794–98.
41. N. Keum et al., "Adult Weight Gain and Adiposity-Related Cancers: A Dose-Response Meta-Analysis of Prospective Observational Studies," *Journal of the National Cancer Institute* 107, no. 2 (March 10, 2015): ii: djv088, doi: 10.1093/jnci/djv088.
42. F. Islami et al., "Proportion and Number of Cancer Cases and Deaths Attributable to Potentially Modifiable Risk Factors in the United States," *CA: A Cancer Journal for Clinicians* 68, 1 (January 2018): 31–54, doi: 10.3322/caac.21440.
43. H. Sung et al., "Emerging Cancer Trends among Young Adults in the USA: Analysis of a Population-based Cancer Registry," *Lancet Public Health* 4, no. 3 (March 1, 2019): https://www.thelancet.com/journals/lanpub/article/PIIS2468-2667(18)30267-6/fulltext.
44. P. Rous, "The Influence of Diet of Transplanted and Spontaneous Mouse Tumors," *Journal of Experimental Medicine* 20, no. 5 (1914): 433–51.

45. A. Tannenbaum, "The Dependence of Tumor Formation on the Composition of the Calorie-Restricted Diet as Well as on the Degree of Restriction," *Cancer Research* 5, no. 11 (1945): 616–25.
46. A. H. Eliassen et al., "Adult Weight Change and Risk of Postmenopausal Breast Cancer," *JAMA* 296, no. 2 (July 12, 2006): 193–201.

Chapter 15: Hyperinsulinemia

1. M. Rabinowitch, "Clinical and Other Observations on Canadian Eskimos in the Eastern Arctic," *Canadian Medical Association Journal* 34 (1936): 487–501.
2. G. M. Brown, L. B. Cronk, and T. J. Boag, "The Occurrence of Cancer in an Eskimo," *Cancer* 5, no. 1 (January 1952): 142–43.
3. G. J. Mouratoff et al., "Diabetes Mellitus in Eskimos," *JAMA* 199, no. 13 (1967): 961–66, doi: 10.1001/jama.1967.03120130047006.
4. George J. Mouratoff et al., "Diabetes Mellitus in Eskimos after a Decade, *JAMA* 226, no. 11 (1973): 1345–46.
5. Cynthia D. Schraer et al., "Prevalence of Diabetes Mellitus in Alaskan Eskimos, Indians, and Aleuts," *Diabetes Care* 11 (1988): 693–700.
6. K. J. Acton et al., "Trends in Diabetes Prevalence among American Indian and Alaska Native Children, Adolescents, and Young Adults," *American Journal of Public Health* 92 (2002): 1485–90.
7. Etan Orgel, "The Links between Insulin Resistance, Diabetes, and Cancer," *Current Diabetes Reports* 13, no. 2 (April 2013): 213–22, doi: 10.1007/s11892-012-0356-6.
8. P. T. Campbell et al., "Diabetes and Cause-Specific Mortality in a Prospective Cohort of One Million U.S. Adults," *Diabetes Care* 35 (2012): 1835–44.
9. S. R. Seshasai et al., "Diabetes Mellitus, Fasting Glucose, and Risk of Cause-Specific Death," *New England Journal of Medicine* 364 (2011): 829–41.
10. Y. Chan et al., "Association between Type 2 Diabetes and Risk of Cancer Mortality: A Pooled Analysis of Over 771,000 Individuals in the Asia Cohort Consortium," *Diabetologia* 60, no. 6 (June 2017): 1022–32, doi: 10.1007/s00125-017-4229-z.
11. T. Stocks et al., "Blood Glucose and Risk of Incident and Fatal Cancer in the Metabolic Syndrome and Cancer Project (Me-Can): Analysis of Six Prospective Cohorts," *PLoS Medicine* 6 (2009): e1000201.
12. E. Giovannucci et al., "Diabetes and Cancer," *Diabetes Care* 33 (2010): 1674–85; S. C. Larsson, N. Orsini, and A. Wolk, "Diabetes Mellitus and Risk of Colorectal Cancer: A Meta-analysis," *Journal of the National Cancer Institute* 97 (2005): 1679–87; S. C. Larsson, C. S. Mantzoros, and A. Wolk, "Diabetes Mellitus and Risk of Breast Cancer: A Meta-analysis," *International Journal of Cancer* 121 (2007): 856–62.
13. W. Wu et al., "Rising Trends in Pancreatic Cancer Incidence and Mortality in 2000–2014," *Clinical Epidemiology* 10 (July 9, 2018): 789–97.

14. B. E. Barker, H. Fanger, and P. Farnes, "Human Mammary Slices in Organ Culture: I. Methods of Culture and Preliminary Observations on the Effects of Insulin," *Experimental Cell Research* 35 (1964): 437–48.
15. D. LeRoith et al., "The Role of Insulin and Insulin-like Growth Factors in the Increased Risk of Cancer in Diabetes," *Rambam Maimonides Medical Journal* 2, no. 2 (2011): e0043.
16. E. J. Gallagher and D. LeRoith, "The Proliferating Role of Insulin and Insulin-like Growth Factors in Cancer," *Trends in Endocrinology and Metabolism* 21, no. 10 (October 2010): 610–18.
17. V. Papa et al., "Elevated Insulin Receptor Content in Human Breast Cancer," *Journal of Clinical Investigations* 86 (1990): 1503–10.
18. J. Ma et al., "A Prospective Study of Plasma C-peptide and Colorectal Cancer Risk in Men," *Journal of the National Cancer Institute* 96 (2004): 546–53.
19. R. Kaaks et al., "Serum C-Peptide, Insulin-like Growth Factor (IGF) I, IGF-Binding Proteins, and Colorectal Cancer Risk in Women," *Journal of the National Cancer Institute* 92, no. 19 (October 4, 2000): 1592–600.
20. E. K. Wei et al., "A Prospective Study of C-peptide, Insulin-like Growth Factor-I, Insulin-like Growth Factor Binding Protein-1, and the Risk of Colorectal Cancer in Women," *Cancer Epidemiology, Biomarkers and Prevention* 14 (2005): 850–55.
21. T. Tsujimoto et al., "Association between Hyperinsulinemia and Increased Risk of Cancer Death in Nonobese and Obese People: A Population-based Observational Study," *International Journal of Cancer* 141 (2017): 102–11.
22. M. J. Gunter et al., "Breast Cancer Risk in Metabolically Healthy but Overweight Postmenopausal Women," *Cancer Research* 75, no. 2 (2015): 270–74.
23. "Three-fold Increase in UK Insulin Use, Study Finds," BBC News, February 6, 2014, https://www.bbc.com/news/health-26065673.
24. C. J. Currie et al., "Mortality and Other Important Diabetes-related Outcomes with Insulin vs Other Antihyperglycemic Therapies in Type 2 Diabetes," *Journal of Clinical Endocrinology and Metabolism* 98, no. 2 (February 2013): 668–77.
25. S. L. Bowker et al., "Increased Cancer-related Mortality for Patients with Type 2 Diabetes Who Use Sulfonylureas or Insulin," *Diabetes Care* 29 (2006): 254–58.
26. C. J. Currie, C. D. Poole, and E. A. M. Gale, "The Influence of Glucose-lowering Therapies on Cancer Risk in Type 2 Diabetes," *Diabetologia* 52 (2009): 1766–77, doi: 10.1007/s00125-009-1440-6.
27. Y. X. Yang, S. Hennessy, and J. D. Lewis, "Insulin Therapy and Colorectal Cancer Risk among Type 2 Diabetes Mellitus Patients," *Gastroenterology* 127 (2004): 1044–50.

Chapter 16: Growth Factors

1. K. B. Michaels and W. C. Willett, "Breast Cancer: Early Life Matters," *New England Journal of Medicine* 351 (2004): 1679–81.
2. P. A. Van den Brandt, "Pooled Analysis of Prospective Cohort Studies on Height, Weight, and Breast Cancer Risk," *American Journal of Epidemiology* 152, no. 6 (September 15, 2000): 514–27.
3. M. Ahlgren et al., "Growth Patterns and the Risk of Breast Cancer in Women," *New England Journal of Medicine* 351 (2004): 1619–26.
4. J. Green et al., "Height and Cancer Incidence in the Million Women Study: Prospective Cohort, and Meta-analysis of Prospective Studies of Height and Total Cancer Risk," *Lancet Oncology* 12, no. 8 (August 2011): 785–94, doi: 10.1016/S1470-2045(11)70154-1.
5. Michelle McDonagh, "Lifestyle Linked to Huge Increase in Short-sightedness," *Irish Times*, February 27, 2018, https://www.irishtimes.com/life-and-style/health-family/lifestyle-linked-to-huge-increase-in-short-sightedness-1.3397726.
6. Elie Dolgin, "The Myopia Boom," *Nature* 519, no. 19 (2015): 276–78.
7. L. C. Cantley, "The Phosphoinositide 3-Kinase Pathway," *Science* 296, no. 5573 (May 31, 2002): 1655–57.
8. Lewis C. Cantley, "Seeking Out the Sweet Spot in Cancer Therapeutics: An Interview with Lewis Cantley," *Disease Models and Mechanisms* 9, no. 9 (September 1, 2016): 911–16, doi: 10.1242/dmm.026856.
9. H. Tan et al., "Genome-wide Mutational Spectra Analysis Reveals Significant Cancer-specific Heterogeneity," *Scientific Reports* 5, no. 12566 (2015): doi: 10.1038/srep12566.
10. L. C. Cantley, "Cancer, Metabolism, Fructose, Artificial Sweeteners and Going Cold Turkey on Sugar," *BMC Biology* 12, no. 8 (2014).
11. M. Barbieri et al., "Insulin/IGF-I-signaling Pathway: An Evolutionarily Conserved Mechanism of Longevity from Yeast to Humans," *American Journal of Physiology-Endocrinology and Metabolism* 285 (2003): E1064–E1071.
12. D. A. Fruman et al., "The PI3K Pathway in Human Disease," *Cell* 170, no. 4 (August 10, 2017): 605–35, doi: 10.1016/j.cell.2017.07.029.
13. Pal A. et al., "PTEN Mutations as a Cause of Constitutive Insulin Sensitivity and Obesity," *New England Journal of Medicine* 367 (2012): 1002–11.
14. D. L. Riegert-Johnson et al., "Cancer and Lhermitte-Duclos Disease Are Common in Cowden Syndrome Patients," *Hereditary Cancer in Clinical Practice* 8, no. 6 (2010): https://doi.org/10.1186/1897-4287-8-6.
15. D. P. Burkitt, "Some Diseases Characteristic of Modern Western Civilization," *BMJ* 1, no. 5848 (February 3, 1973): 274–78, doi: 10.1136/bmj.1.5848.274.
16. Gary Taubes, "Rare Form of Dwarfism Protects Against Cancer," *Discover*, March 26, 2013, http://discovermagazine.com/2013/april/19-double-edged-genes.

17. A. Janecka et al., "Clinical and Molecular Features of Laron Syndrome, a Genetic Disorder Protecting from Cancer," *In Vivo* 30, no. 4 (July–August 2016): 375–81.
18. J. Guevara-Aguirre et al., "Growth Hormone Receptor Deficiency Is Associated with a Major Reduction in Pro-aging Signaling, Cancer, and Diabetes in Humans," *Science Translational Medicine* 3, no. 7 (February 16, 2011): 70ra13, doi: 10.1126/scitranslmed.3001845.
19. J. Jones and D. Clemmons, "Insulin-like Growth Factors and Their Binding Proteins: Biological Actions," *Endocrine Reviews* 16 (1995): 3–34; R. C. Baxter, J. M. Bryson, and J. R. Turtle, "Somatogenic Receptors of Rat Liver: Regulation by Insulin," *Endocrinology* 107, no. 4 (1980): 1176–81; S. J. Moschos and C. S. Mantzoros, "The Role of the IGF System in Cancer: From Basic to Clinical Studies and Clinical Applications," *Oncology* 63 (2002): 317–32; E. Giovannucci and D. Michaud, "The Role of Obesity and Related Metabolic Disturbances in Cancers of the Colon, Prostate, and Pancreas," *Gastroenterology* 132 (2007): 2208–25.
20. M. J. Gunter et al., "A Prospective Evaluation of Insulin and Insulin-like Growth Factor-I as Risk Factors for Endometrial Cancer," *Cancer Epidemiology, Biomarkers and Prevention* 17, no. 4 (2008): 921–29.
21. M. J. Gunter et al., "Insulin, Insulin-like Growth Factor-I, Endogenous Estradiol, and Risk of Colorectal Cancer in Postmenopausal Women," *Cancer Research* 68, no. 1 (2008): 329–37.
22. A. Canonici et al., "Insulin-like Growth Factor-I Receptor, E-cadherin and Alpha-V Integrin Form a Dynamic Complex Under the Control of Alpha-catenin," *International Journal of Cancer* 122 (2008): 572–82.
23. R. Palmqvist et al., "Plasma Insulin-like Growth Factor 1, Insulin-like Growth Factor Binding Protein 3, and Risk of Colorectal Cancer: A Prospective Study in Northern Sweden," *Gut* 50 (2002): 642–46.
24. J. Ma et al., "Prospective Study of Colorectal Cancer Risk in Men and Plasma Levels of Insulin-like Growth Factor (IGF)-1 and IGF-binding Protein-3," *Journal of the National Cancer Institute* 91 (1999): 620–25.

Chapter 17: Nutrient Sensors

1. "Did a Canadian Medical Expedition Lead to the Discovery of an Anti-aging Pill?," *Financial Post*, February 12, 2015, https://business.financialpost.com/news/did-a-canadian-medical-expedition-lead-to-the-discovery-of-an-anti-aging-pill.
2. K. Hara et al., "Amino Acid Sufficiency and mTOR Regulate p70 S6 Kinase and eIF-4E BP1 through a Common Effector Mechanism," *Journal of Biological Chemistry* 273 (1998): 14484–94.
3. B. Magnuson et al., "Regulation and Function of Ribosomal Protein S6 Kinase (S6K) within mTOR Signaling Networks," *The Biochemical Journal* 441, no. 1 (2012): 1–21.
4. "Organ Transplants and Cancer Risk," National Institutes of Health, November 21, 2011, https://www.nih.gov/news-events/nih-research-matters/organ-transplants-cancer-risk.

5. H. Populo et al., "The mTOR Signalling Pathway in Human Cancer," *International Journal of Molecular Science* 13 (2012): 1886–918, doi: 10.3390/ijms13021886.
6. S. A. Forbes et al., "COSMIC: Mining Complete Cancer Genomes in the Catalogue of Somatic Mutations in Cancer," *Nucleic Acids Research* 39 (2011): D945–D950.
7. A. G. Renehan, C. Booth, and C. S. Potten, "What Is Apoptosis, and Why Is It Important?," *British Medical Journal* 322 (2001): 1536–38.
8. Y. H. Tseng et al., "Differential Roles of Insulin Receptor Substrates in the Anti-apoptotic Function of Insulin-like Growth Factor-1 and Insulin," *Journal of Biological Chemistry* 277 (2002): 31601–11.
9. H. Zong et al., "AMP Kinase Is Required for Mitochondrial Biogenesis in Skeletal Muscle in Response to Chronic Energy Deprivation," *Proceedings of the National Academy of Sciences USA* 99 (2002): 15983–87.
10. H. J. Weir et al., "Dietary Restriction and AMPK Increase Life Span via Mitochondrial Network and Peroxisome Remodeling," *Cell Metabolism* 26 (2017): 1–13.

Chapter 18: The Warburg Revival

1. A. M. Otto, "Warburg Effect(s): A Biographical Sketch of Otto Warburg and His Impacts on Tumor Metabolism," *Cancer and Metabolism* 5, no. 5 (2016), doi: 10.1186/s40170-016-0145-9.
2. O. Warburg et al., "Versuche an überlebendem carcinom-gewebe," *Wiener klinische Wochenschrift* 2 (1923): 776–77.
3. O. Warburg, "On the Origin of Cancer," *Science* 123, no. 3191 (1956): 309–14.
4. F. Weinberg et al., "Mitochondrial Metabolism and ROS Generation Are Essential for Kras-mediated Tumorigenicity," *Proceedings of the National Academy of Sciences USA* 107 (2010): 8788–93; A. S. Tan et al., "Mitochondrial Genome Acquisition Restores Respiratory Function and Tumorigenic Potential of Cancer Cells Without Mitochondrial DNA," *Cell Metabolism* 21 (2015): 81–94; V. R. Fantin, J. St-Pierre, and P. Leder, "Attenuation of LDH-A Expression Uncovers a Link between Glycolysis, Mitochondrial Physiology, and Tumor Maintenance," *Cancer Cell* 9 (2006): 425.
5. C-H. Chang et al., "Posttranscriptional Control of T Cell Effector Function by Aerobic Glycolysis," *Cell* 153, no. 6 (2013): 1239–51.
6. X. L. Zu and M. Guppy, "Cancer Metabolism: Facts, Fantasy, and Fiction," *Biochemical and Biophysical Research Communications* 313 (2004): 459–65.
7. W. H. Koppenol et al., "Otto Warburg's Contributions to Current Concepts of Cancer Metabolism," *Nature Reviews Cancer* 11, no. 5 (May 2011): 325–37, doi: 10.1038/nrc3038.
8. M. G. Vander Heiden, "Understanding the Warburg Effect: The Metabolic Requirements of Cell Proliferation," *Science* 324, no. 5930 (May 22, 2009): 1029–33.
9. R. B. Robey and N. Hay, "Is AKT the 'Warburg Kinase'?—AKT: Energy Metabolism Interactions and Oncogenesis," *Seminars in Cancer Biology* 19 (2009): 25–31.

10. R. C. Osthus et al., "Deregulation of Glucose Transporter 1 and Glycolytic Gene Expression by c-Myc," *Journal of Biological Chemistry* 275 (2000): 21797–800.
11. S. Venneti et al., "Glutamine-based PET Imaging Facilitates Enhanced Metabolic Evaluation of Gliomas in Vivo," *Science Translational Medicine* 7, no. 274 (February 11, 2015): 274ra17.
12. H. Eagle, "Nutrition Needs of Mammalian Cells in Tissue Culture," *Science* 122 (1955): 501–14.
13. David R. Wise and Craig B. Thompson, "Glutamine Addiction: A New Therapeutic Target in Cancer," *Trends in Biochemical Sciences* 35, no. 8 (August 2010): 427–33, doi: 10.1016/j.tibs.2010.05.003.
14. A. Carracedo, "Cancer Metabolism: Fatty Acid Oxidation in the Limelight," *Nature Reviews Cancer* 13, no. 4 (April 2013): 227–32.
15. Luana Schito and Gregg L. Semenza, "Hypoxia-inducible Factors: Master Regulators of Cancer Progression," *Trends in Cancer Research* 2, no. 12 (2016): 758–70.
16. G. L. Semenza, "Hypoxia-Inducible Factor 1 and Cancer Pathogenesis," *IUBMB Life* 60, no. 9 (2008): 591–97.
17. G. L. Semenza, "HIF-1 Mediates Metabolic Responses to Intratumoral Hypoxia and Oncogenic Mutations," *Journal of Clinical Investigations* 123 (2013): 3664–71.
18. R. A. Gatenby, "The Potential Role of Transformation-induced Metabolic Changes in Tumor-Host Interaction," *Cancer Research* 55 (1995): 4151–56.
19. V. Estrella et al., "Acidity Generated by the Tumor Microenvironment Drives Local Invasion," *Cancer Research* 73, no. 5 (March 1, 2013): 1524–35, doi: 10.1158/0008-5472.CAN-12-2796.
20. L. Schwartz et al., "Out of Warburg Effect: An Effective Cancer Treatment Targeting the Tumor Specific Metabolism and Dysregulated pH," *Seminars in Cancer Biology*, http://dx.doi.org/doi:10.1016/j.semcancer.2017.01.005.
21. O. Trabold et al., "Lactate and Oxygen Constitute a Fundamental Regulatory Mechanism in Wound Healing," *Wound Repair and Regeneration* 11 (2003): 504–9.

Chapter 19: Invasion and Metastasis

1. National Institutes of Health, s.v., "metastasis," https://www.cancer.gov/publications/dictionaries/cancer-terms/def/metastasis.
2. C. L. Chaffer and R. A. Weinberg, "A Perspective on Cancer Cell Metastasis," *Science* 33, no. 6024 (2011): 1559–64.
3. T. I. Brandler, "Large Fibrolipoma," *British Medical Journal* 1 (1894): 574.
4. V. Estrella, T. Chen, M. Lloyd, et al., "Acidity Generated by the Tumor Microenvironment Drives Local Invasion," *Cancer Research* 73 (2013): 1524–35.
5. A. F. Chambers, A. C. Groom, and I. C. MacDonald, "Dissemination and Growth of Cancer Cells in Metastatic Sites," *Nature Reviews Cancer* 2 (2002): 563–72.
6. C. A. Klein, "Parallel Progression of Primary Tumours and Metastases," *Nature Reviews Cancer* 9 (2009): 302–12, https://doi.org/10.1038/nrc2627.7.

7. H. R. Carlson, "Carcinoma of Unknown Primary: Searching for the Origin of Metastases," *JAAPA* 22, no. 8 (2009): 18–21.
8. S. Meng et al., "Circulating Tumor Cells in Patients with Breast Cancer Dormancy," *Clinical Cancer Research* 10 (2004): 8152–62.
9. S. Nagrath et al., "Isolation of Rare Circulating Tumor Cells in Cancer Patients by Microchip Technology," *Nature* 450 (2007): 1235–39.
10. D. Tarin et al., "Mechanisms of Human Tumor Metastasis Studied in Patients with Peritoneovenous Shunts," *Cancer Research* 44 (1984): 3584–92.
11. S. Braun et al., "A Pooled Analysis of Bone Marrow Micro-Metastasis in Breast Cancer," *New England Journal of Medicine* 353 (2005): 793–802.
12. J. Massagué and A. C. Obenauf, "Metastatic Colonization," *Nature* 529, no. 7586 (January 21, 2016): 298–306, doi: 10.1038/nature17038.
13. D. P. Tabassum and K. Polyak, "Tumorigenesis: It Takes a Village," *Nature Reviews Cancer* 15 (2015): 473–83.
14. D. Hanahan and L. M. Coussens, "Accessories to the Crime: Functions of Cells Recruited to the Tumor Microenvironment," *Cancer Cell* 21 (2012): 309–22.
15. Mi-Young Kim, "Tumor Self-Seeding by Circulating Cancer Cells," *Cell* 139, no. 7 (December 24, 2009): 1315–26, doi: 10.1016/j.cell.2009.11.025.
16. P. K. Brastianos, "Genomic Characterization of Brain Metastases Reveals Branched Evolution and Potential Therapeutic Targets," *Cancer Discovery* 5, no. 11 (November 2015): 1164–77.
17. L. Ding et al., "Genome Remodelling in a Basal-like Breast Cancer Metastasis and Xenograft," *Nature* 464 (2010): 999–1005.

Chapter 21: Cancer Prevention and Screening

1. E. S. Ford et al., "Explaining the Decrease in U.S. Deaths from Coronary Disease, 1980–2000," *New England Journal of Medicine* 356, no. 23 (2007): 2388–98
2. H. K. Weir et al., "Heart Disease and Cancer Deaths: Trends and Projections in the United States, 1969–2020," *Preventing Chronic Disease* 13 (2016): 160211, https://doi.org/10.5888/pcd13.160211.
3. K. G. Hastings et al., "Socioeconomic Differences in the Epidemiologic Transition from Heart Disease to Cancer as the Leading Cause of Death in the United States, 2003 to 2015," *Annals of Internal Medicine* 169, no. 12 (December 18, 2018): 836–44.
4. Stacy Simon, "Facts & Figures 2019: US Cancer Death Rate Has Dropped 27% in 25 Years," American Cancer Society, January 8, 2019, https://www.cancer.org/latest-news/facts-and-figures-2019.html.
5. Anthony Komaroff, "Surgeon General's 1964 Report: Making Smoking History," *Harvard Health Blog*, January 10, 2014, https://www.health.harvard.edu/blog/surgeon-generals-1964-report-making-smoking-history-201401106970.
6. Centers for Disease Control and Prevention, "Smoking Is Down, but Almost 38 Million American Adults Still Smoke," news release, January 18, 2018, https://www.cdc.gov/media/releases/2018/p0118-smoking-rates-declining.html.

7. Islami et al., "Proportion and Number of Cancer Cases and Deaths Attributable to Potentially Modifiable Risk Factors in the United States," 31–54.
8. M. Inoue and S. Tsugane, "Epidemiology of Gastric Cancer in Japan," *Postgraduate Medical Journal* 81 (2005): 419–24.
9. T. Tonda et al., "Detecting a Local Cohort Effect for Cancer Mortality Data Using a Varying Coefficient Model," *Journal of Epidemiology* 25, no. 10 (2015): 639–46, doi: 10.2188/jea.JE20140218.
10. Y. Chen et al., "Excess Body Weight and the Risk of Primary Liver Cancer: An Updated Meta-analysis of Prospective Studies," *European Journal of Cancer* 48, no. 14 (2012): 2137–45.
11. H. C. Taylor and H. B. Guyer, "A Seven-Year History of Early Cervical Cancer," *American Journal of Obstetrics and Gynecology* 52 (1946): 451–55.
12. P. J. Shaw, "The History of Cervical Screening—I: The Pap Test," *Journal of Obstetrics and Gynaecology Canada* 22, no. 2 (2000): 110–14.
13. G. N. Papanicolaou and H. F. Traut, "The Diagnostic Value of Vaginal Smears in Carcinoma of the Uterus," *American Journal of Obstetrics and Gynecology* 42, no. 2 (1941): 193–206.
14. "History of Cancer Screening and Early Detection." American Cancer Society. https://www.cancer.org/cancer/cancer-basics/history-of-cancer/cancer-causes-theories-throughout-history11.html.
15. J. P. Lockhart-Mummery and C. Dukes, "The Precancerous Changes in the Rectum and Colon," *Surgery, Gynecology and Obstetrics* 36 (1927): 591–96.
16. V. A. Gilbertsen and J. M. Nelms, "The Prevention of Invasive Cancer of the Rectum," *Cancer* 41 (1978): 1137–39.
17. S. J. Sinawer, "The History of Colorectal Cancer Screening: A Personal Perspective," *Digestive Diseases and Sciences*, doi: 10.1007/s10620-014-3466-y.
18. S. J. Winawer et al., "Prevention of Colorectal Cancer by Colonoscopic Polypectomy: The National Polyp Study Workgroup," *New England Journal of Medicine* 329 (1993): 1977–81.
19. A. G. Zauber et al., "Colonoscopic Polypectomy and Long-Term Prevention of Colorectal-Cancer Deaths," *New England Journal of Medicine* 366 (2012): 687–96.
20. J. S. Mandel et al., "Reducing Mortality from Colorectal Cancer by Screening for Fecal Occult Blood: Minnesota Colon Cancer Control Study," *New England Journal of Medicine* 328 (1993): 1365–71.
21. Centers for Disease Control and Prevention, "Vital Signs: Colorectal Cancer Screening Test Use—United States, 2012," *Morbidity and Mortality Weekly Report* 62 (2012): 881–88.
22. P. C. Gøtzsche and K. J. Jørgensen. "Screening for Breast Cancer with Mammography," *Cochrane Database Systemic Reviews* 6 (2013): CD001877.
23. N. Biller-Andorno and P. Juni, "Abolishing Mammography Screening Programs? A View from the Swiss Medical Board," *New England Journal of Medicine* 370, no. 21 (May 22, 2014): 1965–67, doi: 10.1056/NEJMp1401875.
24. A. Bleyer and H. G. Welch, "Effect of Three Decades of Screening Mammography on Breast-Cancer Incidence," *New England Journal of Medicine* 367 (2012): 1998–2005.

25. Magnus Løberg et al., "Benefits and Harms of Mammography Screening," *Breast Cancer Research* 17 (2015): 63, doi: 10.1186/s13058-015-0525-z.
26. H. J. Burstein et al., "Ductal Carcinoma In Situ of the Breast," *New England Journal of Medicine* 350, no. 14 (2004): 1430–41, PMID: 15070793.
27. H. D. Nelson et al., "Screening for Breast Cancer: A Systematic Review to Update the 2009 U.S. Preventive Services Task Force Recommendation," Evidence Synthesis No. 124, AHRQ Publication No. 14-05201-EF-1, Agency for Healthcare Research and Quality, Rockville, MD, 2016.
28. Nelson et al., "Screening for Breast Cancer," 16.
29. G. De Angelis et al., "Twenty Years of PSA: From Prostate Antigen to Tumor Marker," *Reviews in Urology* 9, no. 3 (Summer 2007): 113–23.
30. W. J. Catalona et al., "Selection of Optimal Prostate Specific Antigen Cutoffs for Early Detection of Prostate Cancer: Receiver Operating Characteristic Curves," *Journal of Urology* 152, no. 6, part 1 (1994): 2037–42; I. M. Thompson et al., "Prevalence of Prostate Cancer among Men with a Prostate-Specific Antigen Level < or = 4.0 ng per Milliliter," *New England Journal of Medicine* 350 (2004): 2239–46.
31. G. L. Andriole et al. (PLCO Project Team), "Mortality Results from a Randomized Prostate-Cancer Screening Trial," *New England Journal of Medicine* 360, no. 13 (2009): 1310–19.
32. F. H. Schröder et al. (ERSPC Investigators), "Screening and Prostate Cancer Mortality in a Randomized European Study," *New England Journal of Medicine* 360, no. 13 (2009): 1320–28.
33. R. M. Martin et al. (CAP Trial Group), "Effect of a Low-Intensity PSA-based Screening Intervention on Prostate Cancer Mortality: The CAP Randomized Clinical Trial," *JAMA* 319, no. 9 (2018): 883–95.
34. S. Loeb et al., "Complications after Prostate Biopsy: Data from SEER-Medicare," *Journal of Urology* 186 (2011): 1830–34.
35. F. Fang et al., "Immediate Risk of Suicide and Cardiovascular Death After a Prostate Cancer Diagnosis: Cohort Study in the United States," *Journal of the National Cancer Institute* 102 (2010): 307–14.
36. J. J. Fenton et al., "Prostate-Specific Antigen–Based Screening for Prostate Cancer Evidence Report and Systematic Review for the US Preventive Services Task Force," *JAMA* 319, no. 18 (2018): 1914–31.
37. "Final Recommendation Statement. Prostate Cancer: Screening." US Preventative Services Task Force. https://www.uspreventiveservicestaskforce.org/uspstf/recommendation/prostate-cancer-screening.
38. H. S. Ahn, "Korea's Thyroid Cancer 'Epidemic': Screening and Overdiagnosis," *New England Journal of Medicine* 371 (2014): 1765–67.
39. H. R. Harach, K. O. Franssila, and V. M. Wasenius, "Occult Papillary Carcinoma of the Thyroid: A 'Normal' Finding in Finland—A Systematic Autopsy Study," *Cancer* 56 (1985): 531–38.

Chapter 22: Dietary Determinants of Cancer

1. P. T. Scardino, "Early Detection of Prostate Cancer," *Urologic Clinics of North America* 16, no. 4 (November 1989): 635–55.

2. E. T. Thomas et al., "Prevalence of Incidental Breast Cancer and Precursor Lesions in Autopsy Studies: A Systematic Review and Meta-analysis," *BMC Cancer* 17 (2017): 808.
3. J. M. P. Holly, "Cancer as an Endocrine Problem," *Clinical Endocrinology and Metabolism* 22, no. 4 (2008): 539–50.
4. R. Doll and R. Peto, "The Causes of Cancer: Quantitative Estimates of Avoidable Risks of Cancer in the United States Today," *Journal of the National Cancer Institute* 66, no. 6 (1981): 1191–1308.
5. Islami et al., "Proportion and Number of Cancer Cases and Deaths Attributable to Potentially Modifiable Risk Factors in the United States," 31–54; D. M. Parkin, L. Boyd, and L. C. Walker, "The Fraction of Cancer Attributable to Lifestyle and Environmental Factors in the UK in 2010," *British Journal of Cancer* 105, Suppl. 2 (2011): 77s–81s; M. C. Playdon et al., "Weight Gain After Breast Cancer Diagnosis and All-Cause Mortality: Systematic Review and Meta-Analysis," *Journal of the National Cancer Institute* 107, no. 12 (September 30, 2015): djv275, doi: 10.1093/jnci/djv275.
6. M. Arnold et al., "Global Burden of Cancer Attributable to High Body Mass Index in 2012: A Population-Based Study," *Lancet Oncology* 16, no. 1 (2015): 36–46.
7. D. F. Williamson et al., " Prospective Study of Intentional Weight Loss and Mortality in Never-Smoking Overweight US White Women Aged 40–64 Years," *American Journal of Epidemiology* 141 (1995): 1128–41.
8. N. V. Christou et al., " Bariatric Surgery Reduces Cancer Risk in Morbidly Obese Patients," *Surgery for Obesity and Related Disease* 4 (2008): 691–95.
9. L. Sjostrom et al., "Effects of Bariatric Surgery on Cancer Incidence in Obese Patients in Sweden: Swedish Obese Subjects Study," *Lancet Oncology* 10 (2009): 653–62.
10. T. D. Adams et al., "Cancer Incidence and Mortality after Gastric Bypass Surgery," *Obesity* 17 (2009): 796–802.
11. H. Mackenzie et al., "Obesity Surgery and Risk of Cancer," *British Journal of Surgery* 105, no. 12 (November 2018): 1650–57; M. Derogar et al., "Increased Risk of Colorectal Cancer After Obesity Surgery," *Annals of Surgery* 258 (2013): 983–88.
12. P. Kant and M. A. Hull, "Excess Body Weight and Obesity—The Link with Gastrointestinal and Hepatobiliary Cancer," *Nature Reviews Gastroenterology and Hepatology* 8 (2011): 224–38.
13. C. Moreschi, "Beziehungen zwischen Ernährung und Tumorwachstum," *Z Immunitätsforsch, Orig.* 2 (1909): 651–75.
14. V. D. Longo and L. Fontana, "Calorie Restriction and Cancer Prevention: Metabolic and Molecular Mechanisms," *Trends in Pharmacological Sciences* 31, no. 2 (February 2010): 89–98.
15. M. Prisco et al., "Insulin and IGF-1 Receptors Signaling in Protection from Apoptosis," *Hormone and Metabolic Research* 31 (1999): 80–89.
16. M. Kunkel et al., "Overexpression of GLUT-1 and Increased Glucose Metabolism in Tumors Are Associated with a Poor Prognosis in Patients with Oral Squamous Cell Carcinoma," *Cancer* 97 (2003): 1015–24; R. L. Derr et al., "Association between Hyperglycemia and Survival in Patients

with Newly Diagnosed Glioblastoma," *Journal of Clinical Oncology* 27 (2009): 1082–86.
17. C. Yuan et al., "Influence of Dietary Insulin Scores on Survival in Colorectal Cancer Patients," *British Journal of Cancer* 117, no. 7 (2017): 1079–87.
18. Vicente Morales-Oyarvide, "Dietary Insulin Load and Cancer Recurrence and Survival in Patients with Stage III Colon Cancer: Finding from CALGB 89803," *Journal of the National Cancer Institute* 111, no. 2 (2019): 1–10.
19. H. A. Krebs, "The Regulation of the Release of Ketone Bodies by the Liver," *Advances in Enzyme Regulation* 4 (1966): 339–54.
20. W. D. DeWys, "Weight Loss and Nutritional Abnormalities in Cancer Patients: Incidence, Severity and Significance," in K. C. Calman and K. C. H. Fearon, *Clinics in Oncology* (London: Saunders, 1986), 5:251–61.
21. M. J. Tisdale, "Biology of Cachexia," *Journal of the National Cancer Institute* 89 (1997): 1763–73.
22. C. R. Marinac et al., "Prolonged Nightly Fasting and Breast Cancer Risk: Findings from NHANES (2009–2010)," *Cancer Epidemiology, Biomarkers, and Prevention* 24, no. 5 (May 2015): 783–89.
23. C. R. Marinac et al., "Frequency and Circadian Timing of Eating May Influence Biomarkers of Inflammation and Insulin Resistance Associated with Breast Cancer Risk," *PLoS One* 10, no. 8 (2015): e0136240, doi: 10.1371/journal.pone.0136240; Marinac et al., "Prolonged Nightly Fasting and Breast Cancer Risk," 783–89.
24. S. J. Moschos, "The Role of the IGF System in Cancer: From Basic to Clinical Studies and Clinical Applications," *Oncology* 63 (2002): 317–32.
25. Catherine R. Marinac et al., "Prolonged Nightly Fasting and Breast Cancer Prognosis," *JAMA Oncology* 2, no. 8 (August 1, 2016): 1049–55, doi: 10.1001/jamaoncol.2016.0164.
26. F. M. Safdie et al., "Fasting and Cancer Treatment in Humans: A Case Series Report," *Aging* 1, no. 12 (December 31, 2009): 988–1007; T. B. Dorff et al., "Safety and Feasibility of Fasting in Combination with Platinum-based Chemotherapy," *BMC Cancer* 16, 360 (2016).
27. S. de Groot et al., "The Effects of Short-Term Fasting on Tolerance to (Neo) Adjuvant Chemotherapy in HER2-Negative Breast Cancer Patients: A Randomized Pilot Study," *BMC Cancer* 15, 652 (2015).
28. C. Lee et al., "Fasting Cycles Retard Growth of Tumors and Sensitize a Range of Cancer Cell Types to Chemotherapy," *Science Translational Medicine* 4, no. 124 (March 7, 2012): 124ra27.
29. J. M. Evans et al., "Metformin and Reduced Risk of Cancer in Diabetic Patients," *BMJ* 330 (2005): 1304–5; S. L. Bowker et al., "Increased Cancer-Related Mortality for Patients with Type 2 Diabetes Who Use Sulfonylureas or Insulin," *Diabetes Care* 29 (2006): 254–58; G. Libby et al., "New Users of Metformin Are at Low Risk of Incident Cancer: A Cohort Study among People with Type 2 Diabetes," *Diabetes Care* 32 (2009): 1620–25; D. Li et al., "Antidiabetic Therapies Affect Risk of Pancreatic Cancer," *Gastroenterology* 137 (2009): 482–88; G. W. Landman et al., "Metformin Associated with Lower Cancer Mortality in Type 2 Diabetes: ZODIAC-16, *Diabetes Care* 33 (2010): 322–26.

30. M. Bodmer et al., "Long-Term Metformin Use Is Associated with Decreased Risk of Breast Cancer," *Diabetes Care* 33 (2010): 1304–8.
31. P. J. Goodwin et al., "Insulin-Lowering Effects of Metformin in Women with Early Breast Cancer," *Clinical Breast Cancer* 8 (2008): 501–5.
32. S. Yoshizawa et al., "Antitumor Promoting Activity of (-)-epigallocatechin Gallate, the Main Constituent of 'Tannin' in Green Tea," *Phytotherapy Research* 1 (1987): 44–47.
33. I. J. Chen et al., "Therapeutic Effect of High-Dose Green Tea Extract on Weight Reduction: A Randomized, Double-Blind, Placebo-Controlled Clinical Trial," *Clinical Nutrition* 35, no. 3 (June 2016): 592–99, doi: 10.1016/j.clnu.2015.05.003; A. G. Dulloo et al., "Efficacy of a Green Tea Extract Rich in Catechin Polyphenols and Caffeine in Increasing 24-h Energy Expenditure and Fat Oxidation in Humans," *American Journal of Clinical Nutrition* 70, no. 6 (December 1999): 1040–45; S. Rudelle et al., "Effect of a Thermogenic Beverage on 24-Hour Energy Metabolism in Humans," *Obesity* 15 (2007): 349–55.
34. T. Nagao et al., "A Catechin-Rich Beverage Improves Obesity and Blood Glucose Control in Patients with Type 2 Diabetes," *Obesity* 17, no. 2 (February 2009): 310–17, doi: 10.1038/oby.2008.505.
35. P. Bogdanski et al., "Green Tea Extract Reduces Blood Pressure, Inflammatory Biomarkers, and Oxidative Stress and Improves Parameters Associated with Insulin Resistance in Obese, Hypertensive Patients," *Nutrition Research* 32, no. 6 (June 2012): 421–27, doi: 10.1016/j.nutres.2012.05.007.
36. H. Iso et al., "The Relationship between Green Tea and Total Caffeine Intake and Risk for Self-Reported Type 2 Diabetes among Japanese Adults," *Annals of Internal Medicine* 144, no. 8 (April 18, 2006): 554–62.
37. K. Nakachi et al., "Preventive Effects of Drinking Green Tea on Cancer and Cardiovascular Disease: Epidemiological Evidence for Multiple Targeting Prevention," *Biofactors* 13, nos. 1–4 (2000): 49–54.
38. H. Fujiki et al., "Cancer Prevention with Green Tea and Its Principal Constituent, EGCG: From Early Investigations to Current Focus on Human Cancer Stem Cells," *Molecules and Cells* 41, no. 2 (2018): 73–82.
39. M. Shimizu et al., "Green Tea Extracts for the Prevention of Metachronous Colorectal Adenomas: A Pilot Study," *Cancer Epidemiology, Biomarkers, and Prevention* 17 (2008): 3020–25.
40. S. Bettuzzi et al., "Chemoprevention of Human Prostate Cancer by Oral Administration of Green Tea Catechins in Volunteers with High-Grade Prostate Intraepithelial Neoplasia: A Preliminary Report from a One-Year Proof-of-Principle Study," *Cancer Research* 66 (2006): 1234–40.

Chapter 23: Immunotherapy

1. S. A. Hoption Cann, J. P. van Netten, and C. van Netten, "Acute Infections as a Means of Cancer Prevention: Opposing Effects to Chronic Infections? *Cancer Detection and Prevention* 30 (2006): 83–93.

2. S. J. Oiseth et al., "Cancer Immunotherapy: A Brief Review of the History, Possibilities, and Challenges Ahead," *Journal of Cancer Metastasis and Treatment* 3 (2017): 250–61.
3. P. Kucerova and M. Cervinkova, "Spontaneous Regression of Tumour and the Role of Microbial Infection: Possibilities for Cancer Treatment," *Anti-Cancer Drugs* 27 (2016): 269–77.
4. Hoption Cann, van Netten, and van Netten, "Acute Infections as a Means of Cancer Prevention," 83–93.
5. Jerome Groopman, "The T-Cell Army," *New Yorker*, April 16, 2012, https://www.newyorker.com/magazine/2012/04/23/the-t-cell-army.
6. W. B. Coley, "The Treatment of Malignant Tumors by Repeated Inoculations of Erysipelas: With a Report of Ten Original Cases," *American Journal of the Medical Sciences* 105, no. 5 (May 1893): 3–11.
7. P. Ehrlich, "Über den jetzigen Stand der Karzinomforschung," *Ned Tijdschr Geneeskd* 5 (1909): 273–90.
8. F. M. Burnet, "The Concept of Immunological Surveillance," *Progress in Experimental Tumor Research* 13 (1970): 1–27.
9. D. Ribatti, "The Concept of Immune Surveillance Against Tumors: The First Theories," *Oncotarget* 8, no. 4 (2017): 7175–80.
10. L. A. Loeb, "Human Cancers Express Mutator Phenotypes: Origin, Consequences and Targeting," *Nature Reviews Cancer* 11 (2011): 450–57.
11. N. S. Bajaj et al., "Donor Transmission of Malignant Melanoma in a Lung Transplant Recipient 32 Years after Curative Resection," *Transplant Immunology* 23, no. 7 (2010): e26–e31, doi: 10.1111/j.1432-2277.2010.01090.x.
12. R. Pearl, "Cancer and Tuberculosis" *American Journal of Hygiene* 9 (1929): 97–159.
13. A. Morales, D. Eidinger, and A. W. Bruce, "Intercavitary Bacillus Calmette-Guerin in the Treatment of Superficial Bladder Tumors," *Journal of Urology* 116, no. 2 (August 1976): 180–83.
14. A. Morales, "Treatment of Carcinoma In Situ of the Bladder with BCG: A Phase II Trial," *Cancer Immunology, Immunotherapy* 9, nos. 1–2 (1980): 69–72.
15. G. Redelman-Sidi, M. S. Glickman, and B. H. Bochner, "The Mechanism of Action of BCG Therapy for Bladder Cancer: A Current Perspective," *Nature Reviews Urology* 11, no. 3 (March 2014): 153–62.
16. Heidi Ledford, "The Killer Within," *Nature* 508 (2014): 24–26.
17. Charles, Graeber, "Meet the Carousing, Harmonica-Playing Texan Who Won a Nobel for his Cancer Breakthrough," *Wired*, October 22, 2018, https://www.wired.com/story/meet-jim-allison-the-texan-who-just-won-a-nobel-cancer-breakthrough/.
18. D. R. Leach, M. F. Krummel, and J. P. Allison, "Enhancement of Antitumor Immunity by CTLA-4 Blockade," *Science* 271 (1996): 1734–36.
19. D. Schadendorf et al., "Pooled Analysis of Long-Term Survival Data from Phase II and Phase III Trials of Ipilimumab in Unresectable or Metastatic Melanoma," *Journal of Clinical Oncology* 33 (2015): 1889–94.

20. Nobel Assembly at Karolinska Institutet, press release, October 2018, https://www.nobelprize.org/uploads/2018/10/press-medicine2018.pdf.
21. J. D. Wolchok et al., "Overall Survival with Combined Nivolumab and Ipilimumab in Advanced Melanoma," *New England Journal of Medicine* 377 (2017): 1345–56, doi: 10.1056/NEJMoa1709684.
22. "FDA D.I.S.C.O.: First FDA Approval of a CAR T-cell Immunotherapy," Food and Drug Administration, February 23, 2018, https://www.fda.gov/drugs/resources-information-approved-drugs/fda-disco-first-fda-approval-car-t-cell-immunotherapy.
23. M. A. Postow et al., "Immunologic Correlates of the Abscopal Effect in a Patient with Melanoma," *New England Journal of Medicine* 366 (2012): 925–31.
24. R. H. Mole, "Whole Body Irradiation: Radiobiology or Medicine?," *British Journal of Radiology* 26, no. 305 (May 1953): 234–41.
25. G. Ehlers et al., "Abscopal Effect of Radiation in Papillary Adenocarcinoma," *British Journal of Radiology* 46 (1973): 222–24.
26. N. Dagoglu et al., "Abscopal Effect of Radiotherapy in the Immunotherapy Era: Systematic Review of Reported Cases," *Cureus* 11, no. 2 (February 2019): e4103.
27. E. B. Golden et al., "Local Radiotherapy and Granulocyte-Macrophage Colony-Stimulating Factor to Generate Abscopal Responses in Patients with Metastatic Solid Tumours: A Proof-of-Principle Trial," *Lancet Oncology* 16 (2015): 795–803.
28. M. T. Yilmaz et al., "Abscopal Effect, from Myth to Reality: From Radiation Oncologists' Perspective," *Cureus* 11, no. 1 (January 2019): e3860, doi: 10.7759/cureus.3860.
29. C. Vanpouille-Box et al., "DNA Exonuclease Trex1 Regulates Radiotherapy-induced Tumour Immunogenicity," *Nature Communications* 8 (June 9, 2017): 915618, doi: 10.1038/ncomms15618.
30. W. S. M. E. Theelen et al., "Effect of Pembrolizumab After Stereotactic Body Radiotherapy vs Pembrolizumab Alone on Tumor Response in Patients with Advanced Non-Small Cell Lung Cancer: Results of the PEMBRO-RT Phase 2 Randomized Clinical Trial," *JAMA Oncology* 5, no. 9 (July 11, 2019): 1276–82: doi: 10.1001/jamaoncol.2019.1478.
31. R. A. Gatenby, "Population Ecology Issues in Tumor Growth," *Cancer Research* 51, no. 10 (May 15, 1991): 2542–47.
32. Roxanne Khamsi, "A Clever New Strategy for Treating Cancer, Thanks to Darwin," *Wired*, March 25,2019, https://www.wired.com/story/cancer-treatment-darwin-evolution/.
33. J. Zhang et al., "Integrating Evolutionary Dynamics into Treatment of Metastatic Castrate-Resistant Prostate Cancer," *Nature Communications* 8, no. 1 (November 28, 2017): 1816, doi: 10.1038/s41467-017-01968-5.
34. T. Fojo et al., "Unintended Consequences of Expensive Cancer Therapeutics: The Pursuit of Marginal Indications and a Me-Too Mentality that Stifles Innovation and Creativity," *JAMA Otolaryngology—Head and Neck Surgery* 140, no. 12 (2014): 1225–36.

INDEX